# The Practice of Internal Dosimetry in Nuclear Medicine

# Series in Medical Physics and Biomedical Engineering

Series Editors: John G Webster, E Russell Ritenour, Slavik Tabakov, and Kwan-Hoong Ng

Series in Medical Physics and Biomedical Engineering

# The Practice of Internal Dosimetry in Nuclear Medicine

**Michael G. Stabin**
*Vanderbilt University*
*Nashville, Tennessee, USA*

CRC Press
Taylor & Francis Group
Boca Raton  London  New York

CRC Press is an imprint of the
Taylor & Francis Group, an **informa** business

CRC Press
Taylor & Francis Group
6000 Broken Sound Parkway NW, Suite 300
Boca Raton, FL 33487-2742

First issued in paperback 2020

ISBN 13: 978-0-367-57453-6 (pbk)
ISBN 13: 978-1-4822-4581-3 (hbk)

---

**Library of Congress Cataloging-in-Publication Data**

---

Names: Stabin, Michael G., author.
Title: The practice of internal dosimetry in nuclear medicine / Michael G. Stabin.
Other titles: Series in medical physics and biomedical engineering.
Description: Boca Raton, FL : CRC Press, Taylor & Francis Group, [2017] | ©2017 | Series: Series in medical physics and biomedical engineering | Includes bibliographical references and index.
Identifiers: LCCN 2016009690| ISBN 9781482245813 (alk. paper) | ISBN 1482245817 (alk. paper)
Subjects: LCSH: Nuclear medicine. | Radiation dosimetry.
Classification: LCC R895 .S685 2017 | DDC 616.07/57--dc23
LC record available at http://lccn.loc.gov/2016009690

---

**Visit the Taylor & Francis Web site at**
**http://www.taylorandfrancis.com**

**and the CRC Press Web site at**
**http://www.crcpress.com**

# Contents

# Series Preface

The *Series in Medical Physics and Biomedical Engineering* describes the applications of physical sciences, engineering, and mathematics in medicine and clinical research.

The series seeks (but is not restricted to) publications in the following topics:

Artificial Organs
Assistive Technology
Bioinformatics
Bioinstrumentation
Biomaterials
Biomechanics
Biomedical Engineering
Clinical Engineering
Imaging
Implants
Medical Computing and Mathematics
Medical/Surgical Devices
Patient Monitoring
Physiological Measurement
Prosthetics
Radiation Protection, Health Physics, and Dosimetry
Regulatory Issues
Rehabilitation Engineering
Sports Medicine

Systems Physiology
Telemedicine
Tissue Engineering
Treatment

The *Series in Medical Physics and Biomedical Engineering* is the official book series of the International Organization for Medical Physics (IOMP) and an international series that meets the need for up-to-date texts in this rapidly developing field. Books in the series range in level from introductory graduate textbooks and practical handbooks to more advanced expositions of current research.

# The International Organization for Medical Physics

The IOMP represents over 18,000 medical physicists worldwide and has a membership of 80 national and 6 regional organizations, together with a number of corporate members. Membership by default was accorded to individual medical physicists of all national member organizations.

The mission of the IOMP is to advance medical physics practice worldwide by disseminating scientific and technical information, fostering the educational and professional development of medical physics, and promoting the highest quality of medical physics services for patients.

A World Congress on Medical Physics and Biomedical Engineering is held every three years in cooperation with the International Federation for Medical and Biological Engineering (IFMBE) and International Union for Physics and Engineering Sciences in Medicine (IUPESM). A regionally-based international conference, the International Congress of Medical Physics (ICMP) is held between world congresses. IOMP also sponsors international conferences, workshops, and courses.

The IOMP has several programs to assist medical physicists in developing countries. The joint IOMP Library Program supports 75 active libraries in 43 developing countries, and the Used Equipment Donation Program coordinates equipment donations. The Travel Assistance Program provides a limited number of grants to enable physicists to attend the world congresses.

The IOMP co-sponsors the *Journal of Applied Clinical Medical Physics*. It publishes, twice a year, an electronic bulletin, *Medical Physics World*, and it also publishes *e-Zine*, an electronic newsletter about six times a year. IOMP has an agreement with Taylor & Francis for the publication of the *Medical Physics and Biomedical Engineering* series of textbooks, and its members receive a discount.

The IOMP collaborates with international organizations, such as the World Health Organization (WHO), the International Atomic Energy Agency (IAEA), and other international professional bodies such as the International Radiation Protection Association (IRPA) and the International Commission on Radiological Protection (ICRP) to promote the development of medical physics and the safe use of radiation and medical devices.

Guidance on education, training, and professional development of medical physicists is issued by IOMP, which is collaborating with other professional organizations in the development of a professional certification system for medical physicists that can be implemented on a global basis.

The IOMP website (www.iomp.org) contains information on all its activities, policy statements 1 and 2, and "IOMP: Review and Way Forward," which outlines all the activities and plans of IOMP for the future.

# Preface

In 2008, I wrote a book entitled *Fundamentals of Nuclear Medicine Dosimetry*. This was rather a "how to" manual for performing internal dose calculations for application in nuclear medicine. I noted in the preface to this work that there "I reveal practically all of my methods and secrets for practical internal dose calculations." I was not worried about this, particularly; I have plenty of work to do, and am most interested in growth in the area of patient-individualized dosimetry (the subject of Chapter 8 of this book). Hence, even if others learn to do what I do, it is no problem; I will always have enough work. This current book is more descriptive of the current state of the science. In Chapter 1, I go over the mathematical fundamentals again and show a few sample calculations, but the rest of the book describes anthropomorphic models, dosimetric models, and types and uses of diagnostic and therapeutic radiopharmaceuticals. In Chapter 8, I continue my impassioned plea for the nuclear medicine community to treat our nuclear medicine therapy patients with the same high standard of care that our external beam therapy patients enjoy daily. There appears to be hope on the horizon in Europe, and perhaps that will translate into changes in the United States and other countries. There are marvelous drugs on the market that are not being used, and others that are being used in a *one-size-fits-all* method that provides suboptimal therapeutic quality. People could be walking away from many cancers with a more aggressive use of therapeutic radiopharmaceuticals, and there are few side effects during the therapy phase, unlike the severe discomfort, hair loss, etc., in chemotherapy. I remain optimistic that my colleagues will one day be persuaded.

I have been quite blessed to be a part of this exciting field of nuclear medicine dosimetry for over 30 years now. I was honestly surprised when I received an offer from Oak Ridge Associated Universities for my first position out of college with a master's degree in engineering (emphasis on health physics), working at the Radiation Internal Dose Information Center under the tutelage of two of the greats, Evelyn Watson and Roger Cloutier. My fellow students wondered at my choice, since they knew that I could get a much better starting salary in other areas of the nuclear industry. I knew a golden opportunity when I saw it. I spent 15 years learning this field and meeting many fascinating people in the radiopharmaceutical and academic worlds. Today, I look back with gratitude to all of those who have taught me so much and been good friends. I thank God (literally) for all of this, and hope that the material in this text is useful to other professionals in the field.

# Author

**Michael Stabin** is an associate professor in the Department
of Radiology and Radiological Sciences at Vanderbilt University
in Nashville, Tennessee. Before that, he was a visiting
professor at the Universidade Federal de Pernambuco in
Recife, Brazil, for 2 years and was a scientist at the Radiation
Internal Dose Information Center of Oak Ridge Institute
for Science and Education in Oak Ridge, Tennessee, for
15 years. He has a bachelor of science and a master of engineering degree in environmental engineering (health physics
emphasis) from the University of Florida, Gainesville, Florida,
and received his PhD in nuclear engineering (health physics
emphasis) from the University of Tennessee, Knoxville,
Tennessee. He is a Certified Health Physicist (1988, recertified
in 2000, 2004, 2008, and 2012). He is a member of the Health
Physics Society and the Society of Nuclear Medicine. He has
over 150 publications in the open literature, mostly in the
area of internal dosimetry for nuclear medicine applications,
including complete textbooks on health physics and internal
dose assessment. He has served as member and chair of the
American Board of Health Physics Certification Examination
Panel (Part I and II), and as an associate editor of the *Health
Physics Journal* from 1992 until now. He also serves on task
groups of the Society of Nuclear Medicine (the RAdiation

Dose Assessment Resource, or RADAR), American Association of Physicists in Medicine (AAPM), and the International Commission on Radiological Protection (ICRP). He has developed several models, methods, and tools that have become widely used in the nuclear medicine community, including the MIRDOSE and OLINDA/EXM personal computer software codes for internal dose calculations.

## Chapter 1

# Basic Principles of Internal Dosimetry Calculations

Paracelsus, born Philippus Aureolus Theophrastus Bombastus von Hohenheim in 1493, is regarded as the "father of toxicology." He is associated with the quote "Everything is poison, there is poison in everything. Only the dose makes a thing not a poison." In the science of radiation protection, the quantity of central focus is *dose*, which has a very specific definition, to be given shortly. Many radiation effects, be they positive or negative, are generally related to the dose of radiation that a person, animal, or other biological entity has received. Indeed, dose to physical structures can produce radiation damage at high enough levels as well. Thus, radiation protection professionals (health physicists)* focus much of their daily efforts in evaluating radiation doses and their possible consequences.

---

* Paul Frame explores some possible origins of this unusual term at http://www.orau.org/ptp/articlesstories/names.htm#healthphysics, noting that "The term Health Physics originated in the Metallurgical Laboratory at the University of Chicago in 1942, but it is not known exactly why, or by whom, the term was chosen. Most likely, the term was coined by Robert Stone or Arthur Compton."

However, as we will see, this quantity *dose* in radiation protection is often modified by certain factors to account for specific biological responses, and so is an imperfect, although necessary, quantity to calculate or measure.

Radiation doses may be received from external or internal sources. This text is devoted to the evaluation of internal doses (most appropriately called *internal dose assessment* rather than *dosimetry*, as measurements are typically not involved; nonetheless, *internal dosimetry* is the term generally used), and specifically internal doses received due to the practice of nuclear medicine. Internal exposures to radioactive materials also occur in some industrial practices; while all of the principles involved in the calculations are the same, the applications are different and will not be treated in much detail in this text.

## Quantities and Units

The basic quantity of radiation dosimetry is *absorbed dose*, which is just the energy absorbed by any object per unit mass of the object. Absorbed dose is relevant for any kind of radiation being absorbed by any kind of matter. Of course, we are mostly interested in the absorbed dose to human tissue, but one may calculate the absorbed dose to any material (e.g., air, water, wood). The textbook definition of absorbed dose is:

$$\frac{d\varepsilon}{dm} \tag{1.1}$$

where $d\varepsilon$ is the differential energy deposited in mass $dm$. So, for any application, we just need to calculate how much energy is absorbed by an object and divide by its mass. The quantity of absorbed dose is the gray (Gy), which is 1 J/kg. In internal dose calculations, our objects are either normal tissues or tumors, although we may also be interested in dose to different regions of organs (e.g., kidney medulla vs. kidney cortex), and we can even go to the voxel level to calculate doses to very small portions of organs, depending on the resolution of our images. Many radiation effects are well predicted by the simple quantity absorbed dose, but other more complicated quantities are sometimes needed to explain all of the radiation effects we observe; we will describe attempts to characterize these other quantities later in the chapter. But, to begin the discussion, we will describe how to calculate the quantity absorbed dose for internal emitters. A simple equation for the absorbed dose rate in an organ can be shown as (Stabin 2008):

$$\dot{D} = \frac{k \, A \sum_i n_i \, E_i \, \phi_i}{m} \qquad (1.2)$$

where:

$\dot{D}$ is the absorbed dose rate (Gy/s)

$A$ is the activity in the organ (MBq)

$n_i$ is the number of radiations with energy $E_i$ emitted per nuclear transformation

$E_i$ is the energy per radiation (MeV)

$\phi_i$ is the fraction of energy $E_i$ emitted per decay that is absorbed in the organ

$m$ is the mass of the organ (kg)

$k$ is a proportionality constant that expresses the dose in the desired units, given the units employed for the other terms (e.g., Gy – kg/MBq – s – MeV)

For the equation as described here, $k$ would be:

$$k = \left( \frac{10^6 \text{ dis}}{\text{MBq} - \text{s}} \right) \left( \frac{\text{Gy} - \text{kg}}{1 \text{ J}} \right) \left( \frac{1.6 \times 10^{-13} \text{ J}}{\text{MeV}} \right)$$

$$= 1.6 \times 10^{-7} \left( \frac{\text{Gy} - \text{kg}}{\text{MBq} - \text{s} - \text{MeV}} \right) \qquad (1.3)$$

This equation assumes a uniform distribution of activity (MBq/kg) within this organ, with radiations being emitted isotropically. The term $\phi$ accounts for the fact that all the energy emitted in the organ may not be absorbed by the organ. For particulate radiations ($\alpha$ and $\beta$, Auger, conversion electrons, and such), their ranges in matter are typically shorter than the dimensions of any organ, so $\phi = 1.0$. For photons ($\gamma$, x-rays), some of their energy will escape the organ, and $\phi < 1.0$.

We are usually interested in calculating the total dose to an organ, integrated over time (although we will see that dose rate can be an important consideration in explaining some biological effects seen in some therapeutic uses of radiopharmaceuticals). Integration of Equation 1.2 usually only involves integration of the term $A$, as the other terms in the equation do not vary with time (although we will discuss exceptions to this later). For now, we will use the approximation that:

$$\int_0^\infty \dot{D} \, dt = D = \frac{k \sum_i n_i \, E_i \, \phi_i}{m} \int_0^\infty A \, dt \qquad (1.4)$$

Now, $D$ has the unit of Gy. The integral of the activity in an organ over time is simply an expression of the number of disintegrations that have occurred in the organ, as activity is expressed in units of disintegrations per unit time (e.g., 1 MBq = $10^6$ disintegrations/s). This time integral has been given various names; we can use the term *cumulated activity*, which is sometimes shown by the symbol $\tilde{A}$. Therefore, the cumulative dose is given by:

$$D = \frac{k \, \tilde{A} \sum_i n_i \, E_i \, \phi_i}{m} \qquad (1.5)$$

To solve a complete problem, one must sum up the contributions from all source organs (*S*) to target regions of interest (*T*):

$$D(T) = \sum_S \frac{k \, \tilde{A}_S \sum_i n_i \, E_i \, \phi(S \leftarrow T)_i}{m_T} \tag{1.6}$$

Another permutation that one might encounter uses the term *specific absorbed fraction* $\Phi$; this term is the absorbed fraction $\phi$ divided by the target organ mass $m_T$. So $\Phi(S \leftarrow T) = \phi(S \leftarrow T)/m_T$, and the dose equation may be shown as:

$$D(T) = \sum_S k \, \tilde{A}_S \sum_i n_i \, E_i \, \Phi(S \leftarrow T)_i \tag{1.7}$$

## Effective Half-Time Concept

The calculation of the cumulated activity term $\left(\tilde{A}\right)$ requires implementation of the *effective half-time* concept. It is fortunate that most clearance processes from organs and from the body are governed by *first-order* kinetics—that is, the amount of activity as a function of time diminishes in an exponential fashion. Typical time-activity curves take the form of one or more exponential terms (e.g., there may be early, rapid clearance and later, slower clearance), that is, $A(t) = A_1 \, e^{-\lambda_1 t} + A_2 \, e^{-\lambda_2 t} + \cdots.^*$ The rate constants $\lambda_1$ and $\lambda_2$ are rate constants for removal of activity components $A_1$ and $A_2$. Radioactive material also decays by a first-order process, $A(t) = A_0 \, e^{-\lambda_p t}$. The radioactive rate constant $\lambda_p$ can be related to the radionuclide half-life by $\lambda_p = 0.693/T_{1/2}$.

---

\* If activity is not simply continuously decreasing, but has a visible phase of uptake, one of the terms will have a negative exponent (e.g., $A(t) = A_1 \, e^{-\lambda_1 t} - A_2 \, e^{-\lambda_2 t}$).

This is the time needed for half of the radioactive material to decay. Radioactive half-times are characteristic of any given radionuclide. Here are some examples:

| P-32 | 14.3 days | Tc-99 min | 6.02 h |
|------|-----------|-----------|--------|
| Co-60 | 5.3 years | C-11 | 20.3 min |
| H-3 | 12.3 years | O-15 | 122 s |
| Cs-137 | 30.1 years | F-18 | 1.83 h |
| Ra-226 | 1,600 years | U-238 | $4.48 \times 10^9$ years |
| Rn-222 | 3.8 days | U-235 | $7.04 \times 10^8$ years |

In 1930, Rutherford, Chadwick, and Ellis (1930) concluded that the "rate of transformation of an element has been found to be a constant under all conditions." Nuclear processes are not affected by changes in temperature, pressure, or other macroscopic processes. Our problems with disposal of long-lived radionuclides from high-level nuclear waste could be ameliorated if we could change the long half-lives of some species in the waste. The only way to do this is by *transmutation*, that is, by changing the identity of the radionuclide by changing its nuclear makeup (e.g., by absorption of a neutron). A paper appeared in 2008 (Jenkins et al. 2008) purporting to show evidence of minor variations in radioactive decay rates of two radionuclides, which the authors attributed to variations in the distance between the Earth and Sun, and the influence of some kind of solar activity (although they admitted that the mechanism was unclear). Others (Norman et al. 2009) found no such evidence of these variations. For any practical purpose in the radiation sciences, these physical half-lives may be assumed to be constants.

Similarly, we can describe *biological half-times*, which have the same relationship to the terms $\lambda_1$ and $\lambda_2$ given earlier. A biological half-time is the time for half of the material in an organ to be removed by biological removal. When an organ has an uptake of activity that is being removed as stated from the organ by both radioactive decay and biological removal, the *rate constants* add directly, and a single half-time representing both removal terms may be calculated. This is called the *effective half-time*, and is given by the formula:

$$\lambda_{eff} = \lambda_1 + \lambda_p = \frac{0.693}{T_1} + \frac{0.693}{T_p} \qquad (1.8)$$

$$T_{eff} = \frac{T_1 \times T_p}{T_1 + T_p} \qquad (1.9)$$

Now, if we can express the time-dependent activity in an organ by one or more exponential removal functions, including both physical and biological removal, calculation of $\tilde{A}$ is not difficult, as exponential terms are easy to integrate (here we show just one exponential term for simplicity):

$$\tilde{A} = \int_0^t A(t)\, dt = \int_0^t A_1\, e^{-\lambda_{eff1}\, t}\, dt = \frac{A_1}{\lambda_1} \times \left(1 - e^{-\lambda_{eff1}\, t}\right) \qquad (1.10)$$

However, in most cases, we are interested in the integral over all time, and the expression simplifies (note that 1.443 = 1/0.693):

$$\tilde{A} = \int_0^\infty A(t)\, dt = \frac{A_1}{\lambda_{eff1}} = 1.443 \times A_1 \times T_{eff1} \qquad (1.11)$$

Three principles always apply to the effective half-time:

1. The units for both half-times have to be the same.
2. The effective half-time must always be shorter than the smaller of the biological and physical half-times.
3. As one half-time becomes very large relative to the other, the effective half-time converges to the smaller of the two.

Some examples:

| $T_b = 4\,d$ | $T_p = 4\,d$ | $T_{eff} = \dfrac{4 \times 4}{4 + 4} = 2\,d$ |
|---|---|---|
| $T_b = 4\,d$ | $T_p = 10\,d$ | $T_{eff} = \dfrac{4 \times 10}{4 + 10} = 2.86\,d$ |
| $T_b = 4\,d$ | $T_p = 100\,d$ | $T_{eff} = \dfrac{4 \times 100}{4 + 100} = 3.85\,d$ |
| $T_b = 4\,d$ | $T_p = 1{,}000\,d$ | $T_{eff} = \dfrac{4 \times 1{,}000}{4 + 1{,}000} = 3.98\,d$ |

In any real-life dosimetry problem, we will have more than one source organ and more than one target. We only need to implement Equation 1.6 repeatedly to compile all of the contributions. Each organ will have a $\phi$ value for irradiating itself ($\phi\,(1 \leftarrow 1)$) for each kind of radiation, and, if there are photon emissions, some of the radiation leaving organ 1 may be absorbed in another organ—for example, $\phi\,(2 \leftarrow 1)$, that is, organ 1 irradiating organ 2:

$$D_1 = \frac{k\,\tilde{A}_1 \sum_i n_i\,E_i\,\phi_i(1 \leftarrow 1)}{m_1} + \frac{k\,\tilde{A}_2 \sum_i n_i\,E_i\,\phi_i(1 \leftarrow 2)}{m_1} + \cdots$$

$$(1.12)$$

$$D_2 = \frac{k\,\tilde{A}_1 \sum_i n_i\,E_i\,\phi_i(2 \leftarrow 1)}{m_2} + \frac{k\,\tilde{A}_2 \sum_i n_i\,E_i\,\phi_i(2 \leftarrow 2)}{m_2} + \cdots$$

$$(1.13)$$

## Formalized Dosimetry Systems

The generic equation shown in the preceding text has been implemented by several groups, for different purposes. The sometimes-complex equations look very different, and are a source of confusion for many potential users. *All of the systems are completely identical*; they all sum the energy absorbed by organs of the body per unit mass. They use different symbols for the various quantities defined previously, and sometimes apply different assumptions and conditions to the calculations.

## Quimby and Feitelberg

Quimby and Feitelberg (1963) gave a simple formula for the calculation of thyroid doses from $^{131}$I uptakes:

$$D_\beta = 73.8 \ C \ E_\beta \ T \qquad (1.14)$$

All beta emissions with energy $E_\beta$ are assumed to be absorbed in the thyroid, which has activity concentration $C$ (activity/mass), and the thyroid effective half-time is $T$. The factor of 73.8 is the conversion constant $k$ needed to obtain the dose in the proper units. It includes the factor of 1.443 from Equation 1.11.

## International Commission on Radiological Protection

The formal dosimetry system of the International Commission on Radiological Protection (ICRP) is designed to be used in a system of radiation protection for radiation workers, not nuclear medicine patients. The ICRP (1991) uses the symbol "H," which is the *equivalent dose*, the absorbed dose modified by a *radiation weighting factor*, which roughly accounts for the ability of different radiations to cause biological effects. This will be discussed further later in this chapter. They use the rather odd symbol $U$ for the number of disintegrations in a source organ:

$$U_t = \int_0^t A(t) \ dt = \int_0^t A_1 e^{-\lambda_1 t} \ dt = \frac{A_1}{\lambda_1} \times \left(1 - e^{-\lambda_1 t}\right) \qquad (1.15)$$

The time $t$ is 50 years for a radiation worker and 70 years for members of the public. The cumulative dose equivalent (for a radiation worker) is:

$$H_{50}(T) = \sum_{S} U_{50}(S) \times SEE(T \leftarrow S) \qquad (1.16)$$

The term *SEE* is the *specific effective energy*, which contains all of the other factors in Equation 1.6 plus the radiation weighting factor, $w_R$. After calculating the dose to all relevant organs, the ICRP system takes a step further and estimates permissible intakes of different radionuclides per year, based on dose limits for individual organs and the whole body, as well as permissible concentrations of airborne radionuclides, given assumed breathing rates of typical workers. In this text, we will focus on mathematical details and numerical examples related to radiopharmaceutical dosimetry, but we mention this system for completeness. Notably, the ICRP has a task group on internal dosimetry for radiopharmaceuticals that has published a series of compendia with dose estimates for many radiopharmaceuticals (ICRP 1983–2008).

## Medical Internal Dose Committee

The Medical Internal Dose Committee (MIRD) of the Society of Nuclear Medicine (SNM) established the first complete system of internal dose assessment for radiopharmaceuticals. The MIRD "schema" was originally established in 1976 (Loevinger and Berman 1976). The basic outline of the system was followed by numerous publications establishing absorbed fractions for many situations, decay data, dose functions for various source geometries and types, and many other useful applications. The schema was rewritten twice, the most recent being in 2009 (Bolch et al. 2009). The formulation is perhaps the most complex of all of the systems:

$$D(r_T, T_D) = \int_0^{T_D} \dot{D}(r_T, t)\, dt \tag{1.17}$$

$$= \sum_{r_S} \int_0^{T_D} A(r_S, t)\, S(r_T \leftarrow r_S, t)\, dt \tag{1.18}$$

$$S(r_T \leftarrow r_S, t) = \frac{1}{M(r_T, t)} \sum_i E_i\, Y_i\, \phi(r_T \leftarrow r_S, E_i, t) \tag{1.19}$$

In this expression, $T$ indicates a target region and $S$ a source region, the lowercase $t$ indicates time, $T_D$ is the time of integration, and the symbol $Y$ is used for yield, equivalent to $n$, used in the generic equations in the preceding text.

## RAdiation Dose Assessment Resource

To alleviate some of the confusion experienced by many users trying to decipher the ICRP and MIRD dosimetry systems, the RAdiation Dose Assessment Resource (RADAR) task group of the SNM devised a simpler expression of the dose equation, and furthermore implemented it in software and Internet resources, to facilitate users' calculations. The simplest expression of the dose equation is:

$$D_T = \sum_S N_S \times DF(T \leftarrow S) \tag{1.20}$$

$N_S$ is the number of disintegrations occurring in a source region, and $DF$ contains the rest of the terms in Equation 1.6, which is similar to an SEE factor in the ICRP system or the $S$ value in the MIRD system. The DFs are tabulated for available source and target organs for a given radionuclide, so the user mostly does not have to be concerned with all of the details of emission energies, abundances, absorbed fractions, and organ masses. For any given problem, the main task is to calculate the number of disintegrations in all of the source organs, which is done by characterizing the kinetics in those organs and integrating the time-activity curve.

Facilitation of the calculations in computer code was supplied first in the MIRDOSE software code (Stabin 1996), then again in the OLINDA/EXM computer code (Stabin, Sparks, and Crowe 2005). The name of the code was changed, as the MIRD Committee objected to the use of the name of the committee in the code; the idea was implementation of the MIRD technique in a dosimetry code, but the MIRD Committee worried about their name being associated with code they did not write. Both codes are very easy to use; the user simply selects a radionuclide, one of many body models (*phantoms*, to be discussed in Chapter 2), enters values of *N* for any important source organs, and a simple dosimetry report is generated. Table 1.1 shows an example.

## A Brief Example Calculation

A radiopharmaceutical labeled with $^{99m}Tc$ has a 20% uptake in the liver and a biological half-time of 12 h. What will be the number of disintegrations from an administration of 20 MBq? The physical half-life of $^{99m}Tc$ is approximately 6 h.

$$T_{eff} = \frac{6\,h \times 12\,h}{6\,h + 12\,h} = 4\,h$$

$$N = 1.443 \times A_0 \times T_{eff}$$

$$= 1.443 \times (0.02 \times 0.02\ \text{MBq}) \times 4\,h = 23\ \text{MBq} - h \qquad (1.21)$$

$$N = 23\ \text{MBq} - h\left(\frac{10^6\ \text{Bq}}{\text{MBq}}\right)\left(\frac{1\ \text{dis}}{\text{Bq} - s}\right)\left(\frac{3600\ s}{h}\right)$$

$$= 8.3 \times 10^{10}\ \text{dis}$$

**Table 1.1    Sample Table Generated Using OLINDA/EXM Output for Five Different Phantoms**

|  | Gender Average | | | | |
|---|---|---|---|---|---|
|  | Adult | 15-year-olds | 10-year-olds | 5-year-olds | 1-year-olds |
| Adrenals | 1.37E–02 | 1.60E–02 | 2.40E–02 | 3.78E–02 | 6.69E–02 |
| Brain | 3.76E–02 | 3.78E–02 | 4.02E–02 | 4.39E–02 | 5.79E–02 |
| Breasts | 1.01E–02 | 1.09E–02 | — | — | — |
| Esophagus | 1.41E–02 | 1 69E–02 | 2.65E–02 | 4.25E–02 | 7.75E–02 |
| Eyes | 1.16E–02 | 1.34E–02 | 1.94E–02 | 2.98E–02 | 5.06E 02 |
| Gallbladder wall | 1.42E–02 | 1.60E–02 | 2.42E–02 | 3.57E–02 | 6.17E–02 |
| Left colon | 1.34E–02 | 1.45E–02 | 2.34E–02 | 3.64E–02 | 6.33E–02 |
| Small intestine | 1.30E–02 | 1.55E–02 | 2.45E–02 | 3.84E–02 | 6.70E–02 |
| Stomach wall | 1.31E–02 | 1.48E–02 | 2.25E–02 | 3.54E–02 | 6.18E–02 |
| Right colon | 1.30E–02 | 1.46E–02 | 2.25E–02 | 3.54E–02 | 6.08E–02 |
| Rectum | 2.06E–02 | 2.49E–02 | 3.76E–02 | 5.46E–02 | 8.44E–02 |
| Heart wall | 7.28E–02 | 8.95E–02 | 1.65E–01 | 2.24E–01 | 2.95E–01 |
| Kidneys | 1.18E–02 | 1.37E–02 | 2.10E–02 | 3.41E–02 | 6.03E 02 |
| Liver | 2.45E–02 | 2.87E–02 | 4.22E–02 | 6.09E–02 | 1.01E–01 |
| Lungs | 1.99E–02 | 2.44E–02 | 3.73E–02 | 5.88E–02 | 1.11E–01 |
| Ovaries | 1.82E–02 | 2.08E–02 | 2.84E–02 | 5.04E–02 | 8.67E–02 |
| Pancreas | 1.39E–02 | 1.60E–02 | 2.47E–02 | 3.85E–02 | 6.89E–02 |
| Prostate | 1.82E–02 | 2.20E–02 | 5.49E–02 | 4.95E–02 | 6.94E–02 |
| Salivary glands | 1.21E–02 | 1.44E–02 | 2.19E–02 | 3.34E–02 | 5.80E–02 |
| Red marrow | 1.13E–02 | 1.26E–02 | 1.67E–02 | 2.41E–02 | 4.40E–02 |
| Osteogenic cells | 1.00E–02 | 1.09E–02 | 1.37E–02 | 1.81E–02 | 2.98E–02 |
| Spleen | 1.14E–02 | 1.31E–02 | 2.02E–02 | 3.31E–02 | 5.91E–02 |
| Testes | 1.02E–02 | 1.29E–02 | 2.44E–02 | 3.10E–02 | 5.52E–02 |

*(Continued)*

**Table 1.1** *(Continued)*  **Sample Table Generated Using OLINDA/ EXM Output for Five Different Phantoms**

|  | Gender Average | | | | |
|---|---|---|---|---|---|
|  | Adult | 15-year-olds | 10-year-olds | 5-year-olds | 1-year-olds |
| Thymus | 1.46E–02 | 1.70E–02 | 2.49E–02 | 3.91E–02 | 6.84E–02 |
| Thyroid | 1.08E–02 | 1.31E–02 | 2.07E–02 | 3.35E–02 | 6.06E–02 |
| Urinary bladder wall | 1.46E–01 | 1.63E–01 | 2.45E–01 | 3.56E–01 | 5.31E–01 |
| Uterus | 2.59E–02 | 3.24E–02 | 5.27E–02 | 7.52E–02 | 1.13E–01 |
| Total body | 1.25E–02 | 1.38E–02 | 2.21E–02 | 3.50E–02 | 6.33E–02 |
| Effective dose | 1.92E–02 | 2.20E–02 | 3.23E–02 | 4.82E–02 | 8.01E–02 |

Now, what is the dose from the liver to itself, given that $DF$ (liver ← liver) = $3.23 \times 10^{-15}$ Gy/dis?

$$D = N \times DF = 8.3 \times 10^{10} \text{ dis} \times 3.23 \times 10^{-15} \text{ Gy/dis}$$
$$= 2.68 \times 10^{-4} \text{ Gy} = 0.268 \text{ mGy} \qquad (1.22)$$

A complete problem would, of course, have several source regions and many target regions, but this mainly involves repetition of the preceding calculation, calculating all values of $N$, looking up all of the $DF$s, and adding up all of the contributions. This is very repetitive, and the reason why the use of a computer program is clearly more effective than performing all of this work by hand.

# Other Dose Quantities

## *Equivalent Dose*

Absorbed dose is our basic quantity in internal dose assessment. Absorbed doses can explain many radiation effects—for example, for an external exposure, a dose of 3–4 Gy to the

whole body will cause death in about half of an exposed
population of humans in 30 days (this is referred to as the
$LD_{50/30}$). However, absorbed dose alone is not always adequate
to explain all observed radiation effects. As was alluded to in
the preceding text, the first modification that may be made is
multiplying an absorbed dose in Gy by a *radiation weighting
factor* ($w_R$). Radiation weighting factors are related to, but not
the same as, factors called *relative biological effectiveness* factors (RBEs). RBEs are factors determined experimentally, and
relate to a specific experiment, species, radiation type, and
experimental endpoint. If a dose $D'$ of a given radiation type
produces the same biological endpoint in a given experiment
as a dose $D$ of a reference radiation (typically 250 kVp x-rays),
we can define a quantity called the relative biological effectiveness (RBE) as RBE = $D/D'$. Radiation weighting factors are
*operational* quantities assigned to different radiation types,
generally in radiation protection. The values assigned by the
ICRP (1991) are:

| | |
|---|---|
| Alpha particles, heavy ions | 20 |
| x-ray and gamma radiation | 1.0 |
| Electrons | 1.0 |
| Protons, charged pions | 2.0 |
| Neutrons | Varies with neutron energy |

As noted in the preceding text, the usual designation
for *equivalent dose* is $H$. The unit for equivalent dose is the
sievert (Sv):

$$H(\text{Sv}) = D(\text{Gy}) \times w_R$$

## Effective Dose

The ICRP (1979) created a new dosimetric quantity called the
*effective dose*. This concept embodies knowledge about individual organ radiosensitivities to expressing cancer or genetic

effects and assigns dimensionless *tissue weighting factors* ($w_T$) to each organ in proportion to its sensitivity. The effective dose equivalent is numerically equal to the sum of the products of an organ's actual dose equivalent received and its dimensionless weighting factor. Because of the way in which the weighting factors are derived, the effective dose is the dose equivalent which, if uniformly received by the whole body, would theoretically result in the same risk as from the individual organs receiving these different dose equivalents. This quantity relates only to the expression of *stochastic* effects of radiation on populations, and thus two important restrictions apply:

1. The effective dose may NOT be used in situations involving radiation therapy.
2. The effective dose may NOT be applied to any individual person.

The recommended tissue weighting factors have been modified over the years (Table 1.2), as more information is gained about the effects of radiation on exposed populations, most notably the survivors of the Japanese nuclear weapons bombings in World War II.

More details will be given in later chapters; for now, we will study a very simple example. Assume that, in a given situation, the only three organs with a significant equivalent dose are the lungs, liver, and thyroid, which receive 10, 20, and 50 mSv, respectively. The effective dose, using the $w_R$ values from ICRP 60 (ICRP 1991), would be calculated, respectively, as:

$$10 \text{ mSv} \times 0.12 = 1.2 \text{ mSv}$$

$$20 \text{ mSv} \times 0.05 = 1.0 \text{ mSv}$$

$$50 \text{ mSv} \times 0.05 = 2.5 \text{ mSv}$$

The effective dose is 1.2 + 1.0 + 2.5 = 4.7 mSv. This result indicates that the risk from a uniform whole body equivalent dose of 4.7 mSv would theoretically have the same risk as

**Table 1.2   Tissue Weighting Factors Recommended by the ICRP for Calculation of Effective Dose**

| Organ | ICRP 30 | ICRP 60 | ICRP 103 |
|---|---|---|---|
| Gonads | 0.25 | 0.20 | 0.08 |
| Red marrow | 0.12 | 0.12 | 0.12 |
| Colon | | 0.12 | 0.12 |
| Lungs | 0.12 | 0.12 | 0.12 |
| Stomach | | 0.12 | 0.12 |
| Bladder | | 0.05 | 0.04 |
| Breasts | 0.15 | 0.05 | 0.12 |
| Liver | | 0.05 | 0.04 |
| Esophagus | | 0.05 | 0.04 |
| Thyroid | 0.03 | 0.05 | 0.04 |
| Skin | | 0.01 | 0.01 |
| Bone surfaces | 0.03 | 0.01 | 0.01 |
| Salivary glands, brain | | | 0.01 |
| Remainder | 0.30 | 0.05 | 0.12 |

from receiving 10 mSv to the lungs, 20 mSv to the liver, and 50 mSv to the thyroid. This allows the doses to these organs to be expressed as an equivalent whole body equivalent dose, which can be added to assigned external doses to assess the total dose. It also allows comparison of different types of *diagnostic* medical exposures (even from different modalities, such as x-rays and nuclear medicine).

## Biologically Effective Dose and Equivalent Uniform Dose

In radiation therapy, there have been a number of instances in which absorbed doses (D) do not predict observed radiation effects until allowance for the *dose rate* and the possible nonuniformity of the activity distribution are taken into effect. The application of the concepts of *biologically effective dose* (BED) and *equivalent uniform dose* (EUD) have been shown, in some cases, to help explain the response of normal tissues and tumors to the radiation doses received. More detail about these quantities and their utility will be given in later chapters in which specific therapies are discussed. Basically, the BED takes into account the rate at which doses were received by tissues and their capacity to repair any possible damage. One formulation is:

$$BED = D\left(1 + \frac{D\lambda}{(\mu + \lambda)(\alpha/\beta)}\right) \qquad (1.23)$$

Here, $\alpha$ and $\beta$ are the classic radiosensitivity parameters for a given tissue, from cell survival studies, $\lambda$ is the effective dose rate decay constant, and $\mu$ is a repair constant for the tissue. The only term that may be well known for a given application is $\lambda$. The other terms are general quantities from the radiobiology literature, and probably vary considerably for different patients.

The EUD concept assumes that different, nonuniform dose distributions may be considered equivalent if they elicit the same radiobiological response. One formulation of EUD is (Hobbs and Sgouros 2009):

$$EUD = \left(\sum_i v_i \times D_i^{\alpha}\right)^{\frac{1}{\alpha}} \qquad (1.24)$$

Here, $v_i$ is the partial volume receiving absorbed dose $D_i$, and $\alpha$ is a fitting parameter (Henriquez and Castrillon 2011).

## Conclusions

This chapter has laid out the fundamental mathematical basis for internal dose calculations. Quantities and units relevant to practical calculations were provided, with one simple example worked out. We will now explore some of the terms in the dose equation, and explain the models used to derive terms such as $\phi$, organ masses, and others.

# Bibliography

Bolch WE, Eckerman KF, Sgouros G, and Thomas SRJ. 2009. MIRD pamphlet No. 21: a generalized schema for radiopharmaceutical dosimetry–standardization of nomenclature. *J Nucl Med*, 50(3), 477–484.

Henriquez FC and Castrillon SV. 2011. A quality index for equivalent uniform dose. *J Med Phys*, 36(3), 126–132.

Hobbs RF and Sgouros G. 2009. Calculation of biologically effective dose for piecewise defined dose-rate fits. *Med Phys*, 36(3), 904–907.

ICRP (International Commission on Radiological Protection). 1979. *Limits for Intakes of Radionuclides by Workers*. ICRP Publication 30. Pergamon Press, New York.

ICRP (International Commission on Radiological Protection). 1983–2008. *Radiation Dose Estimates for Radiopharmaceuticals*. ICRP Publications 53, 80, and 106, with addenda. Pergamon Press, New York.

ICRP (International Commission on Radiological Protection). 1991. *Recommendations of the International Commission on Radiological Protection*. ICRP Publication 60 Ann. ICRP 21 (1–3). Pergamon Press, New York.

ICRP (International Commission on Radiological Protection). 2007. *The 2007 Recommendations of the International Commission on Radiological Protection*. ICRP Publication 103 Ann. ICRP 37 (2–4). Pergamon Press, New York.

Jenkins JH, Fischbach E, Buncher JB, Gruenwald JG, Krause DE, and Mattes JJ. 2008. Evidence for correlations between nuclear decay rates and earth–sun distance. *Astropart Phys*, 32(1), 42–46.

Loevinger R and Berman M. 1976. *A Revised Schema for Calculating the Absorbed Dose from Biologically Distributed Radionuclides*. MIRD Pamphlet No. 1, Revised. Society of Nuclear Medicine, New York.

Norman EB, Browne E, Shugart HA, Joshi TH, and Firestone RB. 2009. Evidence against correlations between nuclear decay rates and earth–sun *distance*. *Astropar Phys*, 31(2), 135–137.

Quimby E and Feitelberg S. 1963. *Radioactive Isotopes in Medicine and Biology.* Lea and Febiger, Philadelphia.

Rutherford SE, Chadwick J, and Ellis C. 1930. *Radiations from Radioactive Substances.* Cambridge University Press, Cambridge, UK.

Stabin MG. 1996. MIRDOSE: personal computer software for internal dose assessment in nuclear medicine. *J Nucl Med*, 37(3), 538–546.

Stabin MG. 2008. *Fundamentals of Nuclear Medicine Dosimetry.* Springer, New York, NY.

Stabin MG, Sparks RB, and Crowe E. 2005. OLINDA/EXM: the second-generation personal computer software for internal dose assessment in nuclear medicine. *J Nucl Med*, 46, 1023–1027.

# Chapter 2

# Current Anthropomorphic Models for Dosimetry (Phantoms)

Our fundamental dose equation (Equation 1.6) has two modeling parameters that relate to the physical characteristics of the object for which we wish to calculate dose (usually a patient, but it could also be a tumor, an animal, or other *object*). These parameters are the absorbed fraction ($\phi$) and the organ mass ($m$). The parameter $E$ gives the energy released for each disintegration; the absorbed fraction accounts for how much energy was absorbed in the object of mass $m$; and thus we have energy per unit mass—that is, absorbed dose. To estimate the dose, we need a model of the object; this model assigns masses to target regions of interest (e.g., organs, bone marrow), and contains a geometric representation of the object. In many internal dose problems, as the range of particulate radiations (alpha, beta minus, positron) are small compared to the dimensions of human organs, the absorbed fraction for an organ irradiating itself is assumed to be 1.0, and for an organ irradiating other organs is assumed to be 0.0. However, there are cases in which all of the energy emitted in an organ is not absorbed in that organ, and the *self-absorbed fraction* is less than 1.0, and some other region may have a *cross-absorbed fraction* greater than zero. We use the anthropomorphic models to calculate these absorbed fractions (or *specific absorbed fractions*), as explained in Chapter 1.

## Simple Spherical Constructs

The reference *ICRU sphere* (ICRU is the International Commission on Radiological Units and Measurements) is a very general model representing the whole body as a tissue equivalent sphere, prescribed to have a diameter of 30 cm; a composition by mass of 76.2% O, 10.1% H, 11.1% C, and 2.6% N; and a density of 1 g/cm$^3$ (ICRU 1988). This very simple model allows reproducible calculation of average doses throughout the sphere or part of it, and dose at a given depth (e.g., 1 cm, representing the *deep dose equivalent*).

In the first incarnation of ICRP models for internal dosimetry (ICRP is the International Commission on Radiological Protection), the mass of the whole body and of its various organs and tissues were based on the recommendations of ICRP Publication 23 (ICRP 1975), a document that had been assembled using a very large literature base with data on organ and body masses. In this first treatment, ICRP Publication II (ICRP 1959), the whole body was modeled as a sphere of approximately unit density, and all of the individual organs were modeled as spheres also, most of them of approximately unit density, with two exceptions being bone and lung. This publication presented the first comprehensive system of dose assessment and dose limitation for radiation workers, based on metabolic models for many elements of interest to radiation protection, and absorbed fractions and reference organ masses.

## Equation-Based, "Stylized" Phantoms

More comprehensive body models were developed in the 1970s and 1980s. The surfaces of the whole body and of more than 20 internal organs were described by surface equations—spherical, elliptical, conic sections, cutting planes, and other geometrical objects. Much of this important work was done at the Oak Ridge National Laboratory (ORNL). The first comprehensive phantom was developed by Fisher and Snyder (1966, 1967). These modeling attempts were *adopted* by the Medical Internal Radiation Dosimetry (MIRD) Committee of the Society of Nuclear Medicine, and were published as a supplement to the *Journal of Nuclear Medicine* (JNM) (Snyder et al. 1969). The model contained over 20 defined body regions, of three tissue types—*soft* tissue, skeleton, and lung. The masses of the organs were based on the ICRP *Reference Man* (ICRP 1975); this *median* individual was defined via an extensive review of much scientific litera-ture regarding the mass of individuals in Europe and North America—*Reference Man* was 70 kg in mass and 170 cm height. Despite the name, *Reference Man* also had defined *reference* female organs; all were contained in the single phantom definition; however, naturally no cross-irradiation of male and female organs was permitted. Users have been con-fused when obtaining a table of dose estimates for *Reference Man* and finding doses to ovaries and uterus! Via Monte Carlo studies, Snyder, Ford, and Warner published extensive tables of absorbed fractions for this phantom (1978) and, subsequently, *S-factors* (Equation 1.19) (Snyder et al. 1975). This was slightly revised a few years later (Snyder et al. 1978).

Although others (e.g., Hwang, Shoup, and Poston 1976) attempted to develop models to represent reference individuals of other ages, they were problematic, and the set of reference pediatric models that gained acceptance were those of Cristy and Eckerman (1987). Using the same tissue definitions, five additional phantoms were developed representing newborns,

1-year-olds, 5-year-olds, 10-year-olds, and 15-year-olds (Figure 2.1). Again, all of these models contained both male and female organs. The next type of individual of interest for radiation dosimetry addressed was the pregnant woman. Stabin et al. (1995) created an *adult female* model, which was similar in stature to Cristy and Eckerman's 15-year-old model, but which contained only female organs; they then modified the uterus based on ideas developed by Snyder (unpublished), and moved other abdominal organs to accommodate the enlarged uterus in a reasonable manner. They developed models for a 3-month, 6-month, and 0-month pregnant female; in the latter two models, they modeled some other internal structures (namely, a fetal skeleton and a placenta).

All of the Cristy/Eckerman and Stabin et al. adult, pediatric, and pregnant female phantoms were implemented in two widely distributed software codes—the first was called *MIRDOSE* (Stabin 1996), the second was called *OLINDA/EXM* (Stabin, Sparks, and Crowe 2005). These codes greatly facilitated and standardized dose calculations for these *reference* individuals. Equations 1.12 and 1.13 give the complete description of the general method for performing internal dose calculations.

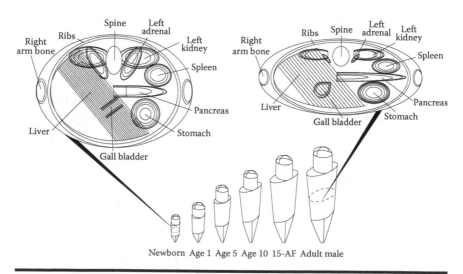

**Figure 2.1   Cristy/Eckerman stylized phantom series.**

As shown, the equation has two source organs, but is depicted as having any number of source and target organs for a given calculation. A typical problem will have half a dozen or so source organs, and generally we want to dose ALL defined target organs; hence, doing a real calculation, by looking up dose factors for every source and target organ, is very tedious and prone to calculation errors. When MIRDOSE was originally written, the use of electronic spreadsheets was not widespread in the sciences. In addition, there are a number of mathematical adjustments—for example, the "remainder of the body correction" (Cloutier et al. 1973a), and calculating the number of disintegrations in the constantly filling and voiding urinary bladder (Cloutier et al. 1973b)—that are not well known to many users. The MIRDOSE and OLINDA/EXM codes provide these features for users.

Stabin (2008) noted how individual organ self-doses for standardized phantoms can be adjusted if organ masses are known to be much different from the models. For electrons, the adjustment is simple:

$$DF_2 = DF_1 \frac{m_1}{m_2} \qquad (2.1)$$

For photons, Snyder (1970) showed that, for self-irradiation, the absorbed fraction increases with the cube root of the mass of the organ, and thus the *specific* absorbed fraction decreases with the 2/3 power of the mass:

$$\phi_2 = \phi_1 \left( \frac{m_2}{m_1} \right)^{\frac{1}{3}}$$

$$\Phi_2 = \Phi_1 \left( \frac{m_1}{m_2} \right)^{\frac{2}{3}} \qquad (2.2)$$

These corrections and other helpful features were included in the OLINDA/EXM software. Thus, these codes were very helpful in producing standardized dose estimates for radio-pharmaceuticals, incorporating the widely accepted ORNL phantom technologies, and providing other established techniques for adjusting dose.

Note, however, that even though this is a text on internal dose calculations for nuclear medicine, all of these phantoms are useful for internal dose calculations for radiation workers as well as external dose calculations in many applications. For example, Markovic, Krstic, and Nikezic (2009) modeled photon and electron doses from radon progeny in the human lung; Takahashi, Shigemori, and Seki (2009) demonstrated a system for radiation dose estimation in emergency response to potential radiological accidents; and the ICRP (2010) provided dose conversion coefficients for a variety of external irradia-tion geometries and particle types. Monte Carlo techniques can model most any type of radiation source and irradiation geometry.

During the testing of some therapeutic radiopharmaceu-ticals, concerns arose about the radiation dose received by the animals involved in the testing. Radiopharmaceutical kinetics observed in animals are routinely used to attempt to estimate doses in humans, although the results are often not particularly reliable (Stabin 2008). However, observed toxicities in some experiments led researchers to wonder what radiation dose the animals themselves might be receiving. Hui et al. (1994) developed a model for an athymic mouse and calculated organ self-doses for $^{90}$Y. Yoriyaz and Stabin (1997) constructed a mouse model and generated dose factors for a selected number of source and target pairs for $^{213}$Bi and $^{90}$Y. Muthuswamy, Roberson, and Buchsbaum (1998) developed a model of marrow for a mouse and provided dose factors for $^{131}$I, $^{186}$Re, and $^{90}$Y. Flynn et al. (2001) developed a mouse model using ellipsoids to define many organs. Konijnenberg et al. (2004) developed a stylized representation of Wistar rats and performed Monte Carlo calculations to develop dose factors for several radionuclides.

## *Image-Based, "Voxel" Phantoms*

In the latter part of the twentieth century, medical imaging techniques permitted the characterization of the internal structures of the human body, with the data stored in digital formats. As 2D medical images are defined by *picture elements*, nicknamed *pixels*, 3D digital representations are called *voxels*, that is, *volume elements*. Xu (2009) described a total of 84 phantoms that various authors developed from CT or MR images of live subjects or cadavers. Considerable effort is needed to manually identify individual voxels as part of identifiable organs or tissues; the phantoms may have hundreds of *slices* and billions of voxels. For medical imaging, the physician routinely interprets the images visually; to digitally assign a large group of voxels as *liver*, for example, one must manually draw regions around every voxel on every slice in which the liver appears and somehow tell the computer that these voxels are part of *liver*, a process called *segmentation*. The volume of individual voxels is determined by their 2D area within the slice and the slice thickness. Possibly the earliest efforts to use medical images for calculation of radiation doses were reported by Gibbs and Pujol (1982). They used 2D x-ray images to form patient images, and used Monte Carlo methods to assess patient doses from medical procedures. Zubal et al. (1994) developed a head and torso model based on CT images of one individual. This model was freely shared with others; others added arms and legs to this phantom (Dawson, Caputa, and Stuchly 1997; Sjogreen et al. 2001, Kramer et al. 2003, 2004) modified organ masses using ICRP-89 reference body heights and organ masses (ICRP 2003) and developed male and female full-body phantoms for use in dosimetry, called MAX and FAX. These were updated to make similar phantoms that were made of polygon mesh surfaces, called MASH and FASH (Kramer et al. 2010). Dimbylow (1996) developed an adult male phantom from MR images that he called *NORMAN*. This phantom was used to determine specific energy absorption rates from exposure to nonionizing electromagnetic fields. He also developed an adult female phantom later, called NAOMI (Dimbylow 2005). Zankl et al. (2005) developed a series of 12 voxel phantoms representing adults and children of various ages from medical images (Williams et al. 1986; Petoussi-Henss et al. 2002; Zanki et al. 2002; Fill et al. 2004). Two of the phantoms developed by this team were adopted by the ICRP as reference computational phantoms (ICRP 2009). Researchers at the University of Florida (Nipper, Williams, and Bolch 2002; Lee et al. 2005, 2006) also developed a series of voxel-based models of adults and children of various ages for proposed use in radiation dosimetry. Xu, Chao, and Bozkurt (2000; Shi and George 2004) developed a phantom based on cross-sectional color photographic images of a human cadaver; they called this phantom *VIP-Man*. It consisted of more than 3.7 billion voxels; the original images were segmented to describe more than 1,400 organs and tissues.

Voxel-based models of animals were also developed for use in radiation dosimetry. Hindorf, Ljungberg, and Strand (2004) developed a model of a mouse using geometric shapes to define several organs but treated the model in a voxel format (Figures 2.2 and 2.3). Kolbert et al. (2003) used MR images of a female athymic mouse to develop realistic models of the kidneys, spleen, and liver, and calculated self-dose and cross-dose S values for the organs. Stabin et al. (2006) manually segmented the micro-CT images of a mouse and rat, and developed specific absorbed fractions for photon and electron sources in the animal organs and dose factors for several source and target regions. Padilla et al. (2008) developed a canine voxel model for use in dosimetry, from a whole-body multislice CT data set.

## *"Hybrid" Phantoms Based on Surface Renderings*

The use of medical image data to represent the human body brought about a second-generational change in the use of anthropomorphic phantoms for radiation dosimetry.

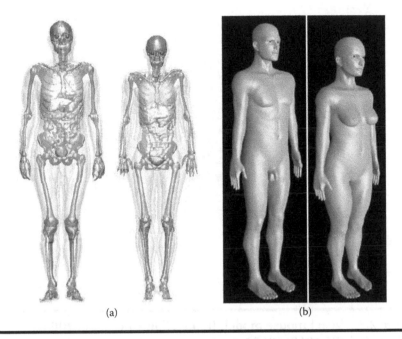

(a)                                              (b)

**Figure 2.2    Realistic phantoms. (a) Voxel phantoms (From Zankl M, et al., 2005, GSF male and female adult voxel models representing ICRP Reference Man—the present status,** *Proceedings of The Monte Carlo Method: Versatility Unbounded in a Dynamic Computing World,* **Chattanooga, TN, American Nuclear Society, A Grange Park, USA. Reproduced with Permission.); (b) MASH and FASH phantoms. (From Kramer R et al., 2010,** *Phys Med Biol,* **55, 163. Reproduced with Permission.)**

The first change, from the use of simple spheres to the use of different objects to represent the various organs and tissues of the body, and allowing calculation of cross-doses, was also profound. Segmentation of medical images to define body organs was very time consuming. Another technology, still based on medical image data, but using surface rendering approaches, facilitated further development of dosimetric phantoms. Segars (2001) developed male and female models using a technique called NonUniform Rational B-Splines (NURBS). Curves and surfaces describe objects using mathematical constructs called *splines.* This technology is used in computer-aided design (CAD) of many manufactured items, and Segars

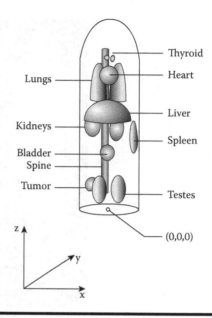

**Figure 2.3   Voxel mouse model. (From Hindorf C et al., 2004,** *J Nucl Med,* **45, 1960–1965.)**

applied it to structures of the human body and the body itself. Once defined, objects can be very easily adjusted and deformed. Segars' NURBS cardiac-torso (NCAT) phantoms are actually 4D models, as the heart deforms while beating, and the lungs and chest move during breathing cycles. Xu and Shi (2005) used the VIP-Man phantom to simulate respiratory motions by adopting the gated respiratory motion data of the NCAT phantom. Then, the 4D VIP-Man phantom was used to study external-beam treatment planning for a lung cancer patient (Zhang et al. 2008). Xu et al. (2007) then developed a series of phantoms representing a pregnant woman and the fetus at 3-, 6-, and 9-months' gestation. These phantoms, which they called the RPI Pregnant Female Phantom Series, were defined by polygonal meshes that were derived from separate anatomical information of a non-pregnant female, a 7-month pregnant woman CT data set, and a mesh model of the fetus. Stabin et al. (2012) designed a series of 12 NCAT models, six male and six female, following the recommended organ masses

from ICRP Publication 89 (ICRP 2003), to replace the six Cristy/
Eckerman (Cristy and Eckerman 1987) reference adult and
pediatric dosimetric models. These, with the three RPI pregnant
female phantoms (which replace those of Stabin et al. 1995),
represent the new generation of reference anthropomorphic
models for use in standardized dosimetry calculations; they are
referred to as the *RADAR* series, as they are associated with the
RAdiation Dose Assessment Resource of the Society of Nuclear
Medicine and Molecular Imaging (SNMMI). In the previous
"generation" of phantoms, the Cristy/Eckerman and Stabin et al.
models were the only completely developed phantoms available
for use. With the advent of NURBS technology, many centers
have developed phantom series rapidly. For example, Tian et al.
(2014) developed a series of 42 pediatric patient models developed using NURBS technology and simulated exposures to various CT exams, and calculated organ doses and effective doses.
In a similar effort, Stabin et al. (2015) developed a series of 92
models, starting with the 12 RADAR models, then adding three
more reference ages, modifying the 50th percentile models to
represent 10th, 25th, 75th, and 90th percentile individuals and
various types of obesity (Figure 2.4). Using a Geant toolkit
model, they also calculated CT doses to all of the phantoms,
for individual organs and effective doses.

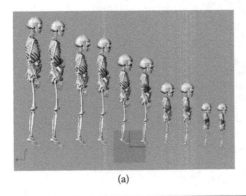
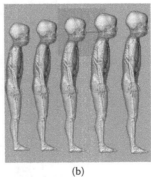

(a)          (b)

**Figure 2.4   RADAR realistic phantoms. (a) Selected RADAR NURBS
phantoms; (b) 10th, 25th, 50th, 75th, and 90th percentile phantoms
for the 5-year-old male NURBS model.**

Keenan et al. (2010) developed a series of three mouse and four rat NCAT models for radiation dosimetry (Figure 2.5) from the one mouse and one rat NCAT model developed by Segars and Tsui (2007). These models were also incorporated into the OLINDA/EXM computer code (Stabin et al. 2005) for routine use in internal dosimetry. Stabin et al. (2015) also developed male and female beagle dog NCAT phantoms for use in dosimetry (Figure 2.6).

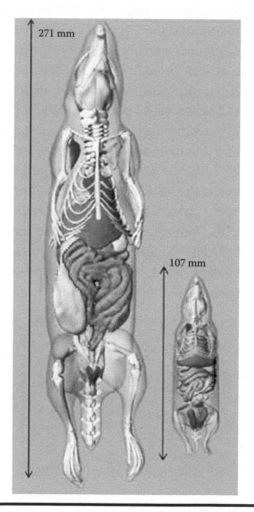

**Figure 2.5** **Images of ROBY and MOBY models used to develop the animal dosimetry models. (From Keenan MA et al., 2010, *J Nucl Med,* 51(3), 471–476.)**

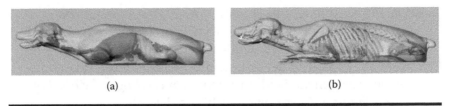
(a)                                    (b)

**Figure 2.6   Female dog model (a) Selected organs; (b) selected skeletal segments. (From Stabin MG et al., 2015, *Health Phys*, 109(3), 198–204.)**

## Summary

The development of human and animal body models progressed very slowly at first, with many years needed to design the models, and to run and check the Monte Carlo results. With the advent of NURBS technology, phantom development has greatly accelerated, and many centers have developed phantoms and phantom series that have many interesting applications, in internal and external dose assessment, for ionizing and nonionizing radiation sources. As very high levels of realism have been achieved and very large numbers of phantoms are available, it is not clear if much more development is required. However, model development will probably be an ongoing endeavor, particularly in the area of skeletal models, which will be discussed in the following chapter.

## Bibliography

Cloutier R, Smith S, Watson E, Snyder W, and Warner G. 1973b. Dose to the fetus from radionuclides in the bladder. *Health Phys*, 25, 147–161.

Cloutier R, Watson E, Rohrer R, and Smith E. 1973a. Calculating the radiation dose to an organ. *J Nucl Med*, 14(1), 53–55.

Cristy M and Eckerman KF. 1987. *Specific Absorbed Fractions of Energy at Various Ages from Internal Photon Sources I: Methods*. ORNL/TM-8381/V1.

Dawson TW, Caputa K, and Stuchly MA. 1997. A comparison of 60 Hz uniform magnetic and electric induction in the human body. *Phys Med Biol*, 42, 2319–2329.

Dimbylow PJ. 1996. The development of realistic voxel phantoms for electromagnetic field dosimetry. *Proceedings of Workshop on Voxel Phantom Development*. Chilton, UK.

Dimbylow PJ. 2005. Resonance behaviour of whole-body averaged specific energy absorption rate (SAR) in the female voxel model, NAOMI. *Phys Med Biol*, 50, 4053–4063.

Fill, UA, Zankl, M, Petoussi-Henss, N. 2004. Adult female voxel models of different stature and photon conversion coefficients for radiation protection. *Health Phys* 86(3), 253–272.

Fisher HLJ and Snyder WS. 1966. *Variation of Dose Delivered by 137Cs as a Function of Body Size from Infancy to Adulthood.* ORNL-4007, 1966.

Fisher HLJ and Snyder WS. 1967. *Distribution of Dose in the Body from a Source of Gamma Rays Distributed Uniformly in an Organ.* ORNL-4168, 1967.

Flynn AA, Green AJ, Pedley RB, Boxer GM, Boden R, and Begent RH. 2001. A mouse model for calculating the absorbed beta-particle dose from $^{131}$I- and $^{90}$Y-labeled immunoconjugates, including a method for dealing with heterogeneity in kidney and tumor. *Radiat Res*, 156, 28–35.

Gibbs SJ and Pujol J. 1982. A Monte Carlo method for patient dosimetry from diagnostic x-ray. *Dentomaxillofac Radiol*, 11, 25.

Hindorf C, Ljungberg M, and Strand S-E. 2004. Evaluation of parameters influencing S values in mouse dosimetry. *J Nucl Med*, 45, 1960–1965.

Hui TE, Fisher DR, Kuhn JA, Williams LE, Nourigat C, Badger CC, Beatty BG, and Beatty JD. 1994. A mouse model for calculating cross-organ beta doses from yttrium-90-labeled immunoconjugates. *Cancer*, 73, 951–957.

Hwang JML, Shoup RL, and Poston JW. 1976. *Mathematical Description of a Newborn Human for Use in Dosimetry Calculations.* ORNL/TM-5453.

ICRP (International Commission on Radiological Protection). 1959. *Report of Committee II on Permissible Dose for Internal Radiation.* ICRP Publication 2, Pergamon Press, New York, NY.

ICRP (International Commission on Radiological Protection). 1975. *Report of the Task Group on Reference Man.* ICRP Publication 23. Pergamon Press, New York, NY.

ICRP (International Commission on Radiological Protection). 2003. *Basic Anatomical and Physiological Data for Use in Radiological Protection: Reference Values*. ICRP Publication 89. Elsevier Health.

ICRP (International Commission on Radiological Protection). 2009. *Adult Reference Computational Phantoms*. ICRP Publication 110. Elsevier Health.

ICRP (International Commission on Radiological Protection). 2010. *Conversion Coefficients for Radiological Protection Quantities for External Radiation Exposures*. ICRP Publication 116. Ann. ICRP, 40(2–5). Pergamon Press, New York, NY.

ICRU (International Commission on Radiological Unit and Measurements). 1988. *ICRU Report 43. Determination of Dose Equivalents from External Radiation Sources-Part 2*. ICRU, Bethesda.

Keenan MA, Stabin MG, Segars WP, and Fernald MJ. 2010. RADAR realistic animal model series for dose assessment. *J Nucl Med*, 51(3), 471–476.

Kolbert KS, Watson T, Matei C, Xu S, Koutcher JA, and Sgouros G. 2003. Murine S factors for liver, spleen and kidney. *J Nucl Med*, 44, 784–791.

Konijnenberg MW, Bijster M, Krenning EP, and de Jong M. 2004. A stylized computational model of the rat for organ dosimetry in support of preclinical evaluations of peptide receptor radionuclide therapy with $^{90}$Y, $^{111}$In, or $^{177}$Lu. *J Nucl Med*, 45, 1260–1269.

Kramer R, Cassola VF, Khoury HJ, Vieira JW, de Melo Lima VJ, and Brown KR. 2010. FASH and MASH: female and male adult human phantoms based on polygon mesh surfaces: II. Dosimetric calculations. *Phys Med Biol*, 55, 163.

Kramer R, Khoury HJ, Vieira JW, Loureiro ECM, Lima VJM, Lima FRA, and Hoff G. 2004. All about FAX: a Female Adult voXel phantom for Monte Carlo calculation in radiation protection dosimetry. *Phys Med Biol*, 49, 5203–5216.

Kramer R, Vieira JW, Khoury HJ, Lima FRA, and Fuelle D. 2003. All about MAX: a male adult voxel phantom for Monte Carlo calculations in radiation protection dosimetry. *Phys Med Biol*, 48, 1239–1262.

Lee C, Lee C, Williams JL, and Bolch WE. 2006. Whole-body voxel phantoms of paediatric patients–UF Series B. *Phys Med Biol*, 51, 4649–4661.

Lee C, Williams JL, Lee C, and Bolch WE. 2005. The UF series of tomographic computational phantoms of pediatric patients. *Med Phys*, 32, 3537–3548.

Markovic VM, Krstic D, and Nikezic D. 2009. Gamma and beta doses in human organs due to radon progeny in human lung. *Radiat Prot Dosim*, 135, 197–202.

Muthuswamy MS, Roberson PL, and Buchsbaum DJ. 1998. A mouse bone marrow dosimetry model. *J Nucl Med*, 39, 1243–1247.

Nipper JC, Williams JL, and Bolch WE. 2002. Creation of two tomographic voxel models of paediatric patients in the first year of life. *Phys Med Biol*, 47, 3143–3164.

Padilla L, Lee C, Milner R, Shahlaee A, and Bolch WE. 2008. Canine anatomic phantom for preclinical dosimetry in internal emitter therapy. *J Nucl Med*, 49, 446–452.

Petoussi-Henss N, Zanki M, Fill U, and Regulla D. 2002. The GSF family of voxel phantoms. *Phys Med Biol*, 47, 89–106.

Segars WP. 2001. Development and Application of the New Dynamic NURBS-based Cardiac-Torso (NCAT) Phantom. PhD Dissertation, The University of North Carolina.

Segars WP and Tsui B. 2007. 4D MOBY and NCAT phantoms for medical imaging simulation of mice and men. *J Nucl Med*, 48(2), 203.

Shi C and George XX. 2004. Development of a 30-week-pregnant female tomographic model from computed tomography (CT) images for Monte Carlo organ dose calculations. *Med Phys*, 31, 2491–2497.

Sjogreen K, Ljungberg M, Wingardh- K, Erlandsson K, and Strand SE. 2001. Registration of emission and transmission whole-body scintillation-camera images. *J Nucl Med*, 42, 1563–1570.

Snyder W. 1970. Estimates of absorbed fraction of energy from photon sources in body organs. In *Medical Radionuclides: Radiation Dose and Effects*, USAEC, pp. 33–50.

Snyder W, Ford M, and Warner G. 1978. *Estimates of Specific Absorbed Fractions for Photon Sources Uniformly Distributed in Various Organs of a Heterogeneous Phantom*. MIRD Pamphlet No. 5, Revised. Society of Nuclear Medicine, New York, NY.

Snyder W, Ford M, Warner G, and Watson S. 1975. *"S," Absorbed Dose Per Unit Cumulated Activity for Selected Radionuclides and Organs*. MIRD Pamphlet No. 11, Society of Nuclear Medicine, New York, NY.

Snyder WS, Fisher HL, Ford MR, and Warner GG. 1969. MIRD Pamphlet No. 5. Estimates of absorbed fractions for monoenergetic photon sources uniformly distributed in various organs of a heterogeneous phantom. *J Nucl Med Suppl*, 3, 7–52.

Stabin M, Watson E, Cristy M, Ryman J, Eckerman K, Davis J, Marshall D, and Gehlen K. 1995. *Mathematical Models and Specific Absorbed Fractions of Photon Energy in the Nonpregnant Adult Female and at the End of Each Trimester of Pregnancy.* ORNL Report ORNL/TM 12907.

Stabin MG. 1996. MIRDOSE: personal computer software for internal dose assessment in nuclear medicine. *J Nucl Med*, 37(3), 538–546.

Stabin MG. 2008. *Fundamentals of Nuclear Medicine Dosimetry.* Springer, New York, NY.

Stabin MG, Kost SD, Segars WP, and Guilmette RA. 2015. Two realistic beagle models for dose assessment. *Health Phys*, 109(3), 198–204.

Stabin MG, Peterson TE, Holburn GE, and Emmons MA. 2006. Voxel-based mouse and rat models for internal dose calculations. *J Nucl Med*, 47, 655–659.

Stabin MG, Sparks RB, and Crowe E. 2005. OLINDA/EXM: The second-generation personal computer software for internal dose assessment in nuclear medicine. *J Nucl Med*, 46, 1023–1027.

Stabin MG, Xu XG, Emmons MA, Segars WP, Shi C, and Fernald MJ. 2012. RADAR Reference Adult, Pediatric, and Pregnant Female Phantom Series for Internal and External Dosimetry. *J Nucl Med*, 53(11), 1807–1813.

Takahashi F, Shigemori Y, and Seki A. 2009. Accurate dose assessment system for an exposed person utilizing radiation transport calculation codes in emergency response to a radiological accident. *Radiat Prot Dosim*, 133, 35–43.

Tian X, Li X, Segars WP, Paulson EK, Frush DP, and Samei E. 2014. Pediatric chest and abdominopelvic CT: organ dose estimation based on 42 patient models. *Radiology*, 270(2), 535–547.

Williams G, Zankl M, Abmayr W, Veit R, and Drexler G. 1986. The calculation of dose from external photon exposures using reference and realistic human phantoms and Monte Carlo methods. *Phys Med Biol*, 31, 449–452.

Xu XG. 2009. Computational phantoms for radiation dosimetry: a 40-year history of evolution. In *Handbook of Anatomical Models for Radiation Dosimetry*, edited by XG Xu and KF Eckerman (Chapter 1). Taylor and Francis, London.

Xu XG, Chao TC, and Bozkurt A. 2000. VIP-Man: an image-based whole-body adult male model constructed from color photographs of the visible human project for multi-particle Monte Carlo calculations. *Health Phys*, 78, 476–486.

Xu XG and Shi CY. 2005. Preliminary development of a 4D anatomical model for Monte Carlo simulations. *The Monte Carlo Method: Versatility Unbounded in a Dynamic Computing World*. American Nuclear Society Proceedings (ISBN: 0-89448-695-0).

Xu XG, Taranenko V, Zhang J, and Shi C. 2007. A boundary-representation method for designing whole-body. Radiation dosimetry models: pregnant females at the ends of three gestational periods—RPI-P3, -P6 and -P9. *Phys Med Biol*. 52, 7023–7044.

Yoriyaz H and Stabin M. 1997. Electron and photon transport in a model of a 30 g mouse [abstract]. *J Nucl Med*, 38(suppl), 228P.

Zanki M, Fill U, Petoussi-Henss N, and Regulla D. 2002. Organ dose conversion coefficients for external photon irradiation of male and female voxel models. *Phys Med Biol*, 47, 2367–2385.

Zankl M, Becker J, Fill U, Petoussi-Henss N, and Eckerman KF. 2005. GSF male and female adult voxel models representing ICRP Reference Man—the present status. *Proceedings of The Monte Carlo Method: Versatility Unbounded in a Dynamic Computing World*. Chattanooga, TN, American Nuclear Society, A Grange Park, USA.

Zankl M, Veit R, Williams G, Schneider K, Fendel H, Petoussi N, and Drexler G. 1988. The construction of computer tomographic phantoms and their application in radiology and radiation protection. *Radiat Environ Biophys*, 27, 153–164.

Zhang JY, Xu GX, Shi C, and Fuss M. 2008. Development of a geometry-based respiratory motion-simulating patient model for radiation treatment dosimetry. *J Appl Clin Med Phys*, 9, 2700.

Zubal IG, Harrell CR, Smith EO, Rattner Z, Gindi G and Hoffer PB. 1994. Computerized 3-Dimensional Segmented Human Anatomy. *Med Phys* 21(2), 299–302.

# Chapter 3

# Use of Animal Models in Internal Dose Calculations

New Drug Applications (NDA) to the US Food and Drug Administration (FDA) require a preclinical phase of testing (i.e., using animal species) and three clinical phases of testing (Stabin 2008). The focus of the testing is to establish a new drug's safety and efficacy; one part of the safety evaluation is the estimation of radiation doses to organs of interest in the human body. Some employ the *first-in-humans* testing method (e.g., Collier et al. 2014), in which testing with cautious levels of dosage are performed in a limited number of human subjects, to show that the compound has the basic desired biodistribution and effects. Some have spent considerable resources in preclinical trials, and, when moving into Phase I clinical trials, very different biodistribution data are seen in humans than was observed in animals. Most preclinical research is done using rodent species, but any animal species in theory is acceptable. Some have an inclination to use primate species, with the idea that they may produce results more similar to humans. Extrapolation of animal data to humans has produced misleading information in many cases, in most any animal species (Stabin 2008). Nevertheless, a preclinical study is a necessary step in the process of evaluating the dosimetry of a new radiopharmaceutical; one must bear in mind that the real dosimetry will not be known until well-designed and executed studies using human subjects are performed.

# Quantitative Methods

Time–activity curves for radiopharmaceuticals using animals may be established by administering the radiopharmaceutical and either sacrificing the animals, extracting tissue samples and performing a radioassay, or by small animal imaging studies.

## Tissue Extraction Methods

Using a minimum of three animals per time point, individual samples of organs and tissues may be extracted from the animals after sacrifice and counted in any radiation detector system (e.g., sodium iodide scintillation, liquid scintillation). Collection of urine and/or fecal samples via use of *metabolic cages* may characterize the excretion of the agent. Extrapolating the organ/tissue data to humans is not an exact science. Extrapolation methods will be discussed in the following text.

## Small Animal Imaging

Small animal imaging techniques have greatly improved the science of drug development. It has also allowed the characterization of radiopharmaceutical dosimetry in living animals, eliminating the need to sacrifice them. A drawback, however, is that the animals generally need to be anesthetized for the imaging session. Anesthetics may alter the distribution of the radiopharmaceutical, leading to inaccurate evaluation of organ uptakes and subsequent dosimetric analyses. Quantitative analyses of organ uptake at any imaging time are the same as those from human imaging studies. PET images are inherently quantitative; counts in a given voxel are easily related to absolute values of activity. Drawing volumes of interest (VOIs) over recognized organ regions provide values of activity in the organ that can be related to percentages or fractions of the administered activity, which are always known. Thus, time–activity curves can be readily (but not easily!) developed from image data and integrated to obtain the time–activity integrals needed for the development of dose estimates. Organ and whole-body biokinetics, combined with analyses of excretion, will allow development of a complete dosimetric analysis.

## Extrapolation Methods

Extrapolating the organ/tissue data to humans is not an exact science. One may assume that the percentage of the administered activity seen in any organ at a given time will likely be of the same concentration seen in humans; one may say that this is a *direct extrapolation.* One may assume that the percentage of administered activity per gram in an organ will be the same in humans; due to the normally considerable differences in body and organ masses, this is likely to produce erroneous results. Crawford and Richmond (1981) and Wegst (1981) evaluated various extrapolation methods proposed in the literature. One method of extrapolating animal data that has been applied by many is the % kg/g method (Kirschner, Ice, and Beierwaltes 1975). In this method, the animal organ data need to be reported as percentage of injected activity per gram of tissue, and this information plus knowledge of the animal total body weight (TB weight) are employed in the following extrapolation:

$$\left(\frac{\%}{organ}\right)_{human} = \left[\left(\frac{\%}{g_{organ}}\right)_{animal} \times \left(kg_{TBweight}\right)_{animal}\right]$$

$$\times \left(\frac{g_{organ}}{kg_{TBweight}}\right)_{human} \quad (3.1)$$

The percentage uptake per gram of tissue is multiplied by the animal whole-body weight in kilograms; the percentage in any organ in humans is obtained by applying the corresponding organ and body masses of a reference adult human. A numerical example using this method has been provided by Stabin (2008).

The animal total body weight was 20 g (0.02 kg), and the source organ chosen had a mass of around 299 g in a human. The human total body weight for the standard adult male of 70 kg was then applied in the transformation. For example (Table 3.1):

$$\frac{38.1\%}{g}(animal) \times 0.020 \text{ kg} \times \frac{299 \text{ g}}{70 \text{ kg}} = \frac{3.26\%}{organ}(human) \quad (3.2)$$

**Table 3.1 Animal Data Extrapolation Example (Mass Extrapolation)**

|  | 1 h | 3 h | 6 h | 16 h | 24 h |
|---|---|---|---|---|---|
| **Animal** | | | | | |
| %ID/organ | 3.79 | 3.55 | 2.82 | 1.02 | 0.585 |
| %ID/g | 38.1 | 36.6 | 30.8 | 11.3 | 5.7 |
| **Human** | | | | | |
| %ID/organ | 3.26 | 3.12 | 2.63 | 0.962 | 0.486 |

The idea of this method is that the percentage in an organ is weighted for the fraction of total body mass that the organ comprises. As noted in the preceding text, this is not a *gold standard* method by any means; it is an attempt to perform a reasonable extrapolation that avoids some of the pitfalls present in the other methods. Some also suggest adding a scaling in time, to account for the different metabolic rates of species of different sizes. Figure 3.1, taken from Glazier (2015), shows that metabolic rates vary with body size in mammals, with slopes typically between 0.67 and 0.75. From this, one may scale the times at which the same metabolic removal may have occurred in two mammalian species:

$$t_{human} = t_{animal} \left[ \frac{m_{human}}{m_{animal}} \right]^{b} \tag{3.3}$$

where $t_{animal}$ is the time at which a measurement was made in an animal system, $t_{human}$ is the corresponding time assumed for the human data, $m_{animal}$ and $m_{animal}$ are the total body masses of the animal species and of the human, and $b$ is

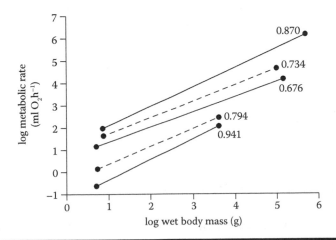

**Figure 3.1** **Variation of metabolic rates with mammalian body size. (From Glazier DS, 2008, *Proc R Soc B*, 275, 1405–1410.)**

**Table 3.2  Animal Data Extrapolation Example (Time Extrapolation)**

| Animal time scale | 5 min | 15 min | 30 min | 60 min | 1.5 h |
|---|---|---|---|---|---|
| Extrapolated human time scale | 22 min | 1.1 h | 2.2 h | 4.3 h | 6.5 h |

a scaling factor derived from the slope of the metabolic rate ratios. Again from Stabin (2008), Table 3.2 shows an example case with data extrapolated from an animal species to a human using this time-scaling approach.

$$t_{human} = 22 \text{ min} = t_{animal} \left[ \frac{m_{human}}{m_{animal}} \right]^{0.25} = 5 \text{ min} \times \left[ \frac{70}{0.2} \right]^{0.25} \quad (3.4)$$

Here, a human body mass of 70 kg was used, and the animal whole-body mass was assumed to be 200 g. Sparks and Aydogan (1999) studied the success of animal data extrapolation for several radiopharmaceuticals, using direct extrapolation and mass and/or time extrapolation. They found that no particular method was superior to any other, and that, in many cases, extrapolated animal data significantly underestimated observed uptakes in human organs (Figure 3.2). Hence, in conclusion, choice of an animal species and extrapolation method are areas of freedom in designing an animal study, and results obtained from animal studies must be recognized as only preliminary estimates of the dosimetry for any radiopharmaceutical.

## Calculating Dose to Animals

Chapter 2 summarized efforts to develop dosimetric models for animals. The most common use of preclinical studies is to provide data that can be extrapolated to humans in order to estimate dose in human organs prior to the start of clinical studies.

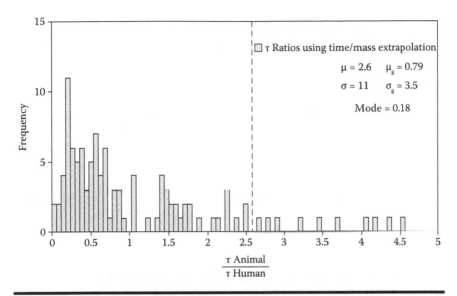

**Figure 3.2  Frequency distribution of the ratio of organ time–activity integrals in animals and humans when both time and mass extrapolations were applied to the animal data. (From Sparks RB and Aydogan B, 1999, In *Sixth International Radiopharmaceutical Dosimetry Symposium*, Oak Ridge Institute for Science and Education, Oak Ridge, TN, pp. 705–716.)**

Particularly with potential therapeutic agents, some have become interested in calculating dose to the animal organs too, in case radiation-related toxicity might be observed.

## *Theoretical Example*

For a human subject, we might administer 500 MBq of a substance labeled with $^{131}$I, for which 30% is taken up by the liver and has an effective half-time of 10 h. The number of disintegrations would be:

$$N = 1.443 \times 500 \text{ MBq} \times 0.3 \times 10 \text{ h} \times \frac{3,600 \text{ s}}{\text{h}}$$

$$= 7.79 \times 10^6 \text{ MBq} - \text{s} \tag{3.5}$$

The dose to the liver and marrow, based on the dose factors in OLINDA/EXM 1.1 (Stabin, Sparks, and Crowe 2005), would be:

$$D_{liver} = 7.79 \times 10^6 \, MBq - s \times 2.86 \times 10^{-5} \, \frac{mGy}{MBq - s} = 223 \, mGy$$

$$(3.6)$$

$$D_{marrow} = 7.79 \times 10^6 \, MBq - s \times 2.98 \times 10^{-7} \, \frac{mGy}{MBq - s} = 2.32 \, mGy$$

$$(3.7)$$

Now consider administering this activity to a 400 g rat. The animal is about 400/70,000 times smaller than an adult human. Adjusting the administered activity by this ratio, we might give only 3 MBq of this substance to the rat. The cumulated activity and dose to the liver would be (using dose factors for the rat from Keenan et al. 2010):

$$N = 1.443 \times 3 \, MBq \times 0.3 \times 10 \, h \times \frac{3,600 \, s}{h}$$

$$= 4.6 \times 10^4 \, MBq - s \qquad (3.8)$$

$$D_{liver} = 4.6 \times 10^4 \, MBq - s \times 2.06 \times 10^{-3} \, \frac{mGy}{MBq - s}$$

$$= 96 \, mGy \qquad (3.9)$$

Entering data for all organs for [18]FDG from the ICRP-recommended biokinetic model (ICRP 2008) for both an adult human and a 400 g rat, we can make a comparison of the radiation dose per MBq that our models predict to selected organs (Table 3.3).

**Table 3.3  Theoretical Radiation Dose Estimates (mGy/MBq) to a 400 g Rat and 70 kg Human from $^{18}$FDG**

| Target Organ | 400 g Rat | 70 kg Human |
|---|---|---|
| Brain | 3.67E–01 | 3.56E–02 |
| Large intestine | 2.07E–01 | 1.13E–02 |
| Small intestine | 1.83E–01 | 1.22E–02 |
| Stomach wall | 1.96E–01 | 1.24E–02 |
| Heart | 4.71E–01 | 6.34E–02 |
| Kidneys | 1.79E–01 | 1.07E–02 |
| Liver | 2.56E–01 | 2.19E–02 |
| Lungs | 3.42E–01 | 1.76E–02 |
| Pancreas | 1.95E–01 | 1.26E–02 |
| Skeleton | 7.26E–01 | 1.04E–02 |
| Spleen | 1.70E–01 | 1.01E–02 |
| Testes | 5.72E–02 | 1.02E–02 |
| Thyroid | 9.44E–02 | 1.01E–02 |
| Urinary bladder | 1.64E+00 | 1.36E–01 |
| Total Body | 1.45E–01 | 1.09E–02 |

## Literature Example

Johnson et al. (2011) injected mice with $^{99m}$Tc methylene diphosphonate (MDP) and $^{18}$F (fluoride) to evaluate image quality in small animal SPECT and PET imaging, respectively (Figure 3.3). Large osteoblastic tumors were well visualized with either modality, but some smaller tumors were better visualized by SPECT imaging. Using the small animal models of Keenan et al. (2010), they found that either agent gave a radiation dose of about 590 mGy to the skeleton.

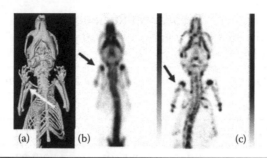

**Figure 3.3** **Humeral lesion is seen in the CT volume rendering (a) as a region lacking bone, and subsequently the corresponding cold spot can be seen in SPECT (c) but not PET (b). (From Johnson LC et al.,** *Radiation Dose-based Comparison of PET and SPECT for Preclinical Bone Imaging,* **2011 IEEE Nuclear Science Symposium Conference Record © 2011 IEEE. http://ieeexplore.ieee.org/stamp/stamp.jsp?tp=arnumber=6152502.)**

# Bibliography

Collier T, Yokel D, Rice P, Jackson R, Shoup T, Normandin M, Brady T, El Fahhri G, Liang S, and Vasdev N. 2014. First human use of the tau radiopharmaceutical [18F]T807 by microfluidic flow chemistry. *J Nucl Med*, 55(1), 1246.

Crawford DJ and Richmond CR. 1981. Epistemological considerations in the extrapolation of metabolic data from non-humans to humans. In *Third International Radiopharmaceutical Dosimetry Symposium*, edited by E Watson, A Schlafke-Stelson, J Coffey and R Cloutier (pp. 191–197). US Department of Health, Education, and Welfare, Oak Ridge, TN.

Glazier DS. 2008. Effects of metabolic level on the body size scaling of metabolic rate in birds and mammals. *Proc R Soc B*, 275, 1405–1410.

ICRP (International Commission on Radiological Protection). 2008. *Radiation Dose to Patients from Radiopharmaceuticals— Addendum 3 to ICRP Publication 53.* ICRP Publication 106. Ann. ICRP 38(1–2). Pergamon Press, New York, NY.

Johnson LC, Johnson RW, Sterling JA, Stabing MG, and Peterson TE. 2011. *Radiation Dose-based Comparison of PET and SPECT for Preclinical Bone Imaging.* 2011 IEEE Nuclear Science Symposium Conference Record, http://ieeexplore.ieee.org/stamp/stamp.jsp?tp=arnumber=6152502.

Keenan MA, Stabin MG, Segars WP, and Fernald MJ. 2010. RADAR realistic animal model series for dose assessment. *J Nucl Med*, 51(3), 471–476.

Kirschner A, Ice R, and Beierwaltes W. 1975. Radiation dosimetry of 131I-19-iodocholesterol: the pitfalls of using tissue concentration data, the author's reply. *J Nucl Med*, 16(3), 248–249.

Sparks RB and Aydogan B. 1999. Comparison of the effectiveness of some common animal data scaling techniques in estimating human radiation dose. In *Sixth International Radiopharmaceutical Dosimetry Symposium*, edited by A Stelson, M Stabin and R Sparks (pp. 705–716). Oak Ridge Institute for Science and Education, Oak Ridge, TN.

Stabin MG. 2008. *Fundamentals of Nuclear Medicine Dosimetry*. Springer, New York, NY.

Stabin MG, Sparks RB, and Crowe E. 2005. OLINDA/EXM: the second-generation personal computer software for internal dose assessment in nuclear medicine. *J Nucl Med*, 46, 1023–1027.

Wegst A. 1981. Collection and presentation of animal data relating to internally distributed radionuclides. In *Third International Radiopharmaceutical Dosimetry Symposium*, edited by E Watson, A Schlafke-Stelson, J Coffey and R Cloutier (October 7–10). US Department of Health, Oak Ridge, Tennessee.

# Chapter 4

# Special Dosimetry Models

Internal dose calculations require a series of modeling assumptions. In Chapter 1, we reviewed the calculational steps, and in Chapter 2, we looked at standard anthropomorphic models (phantoms) used in the calculations. These anthropomorphic models contain regions representing the major organs of the human (or animal!) body. In addition to these organs, there are other organs and tissues that are of interest for internal dosimetry; we will review some of the models for treating those organs in this chapter.

## Bone and Marrow Models

Calculating radiation doses to tissues of the skeletal system is much more difficult than for many other tissues, owing to the complex geometries involved and the intimate mixing of tissues of different densities and atomic compositions. The liver is comprised of several lobes, but the whole liver is treated as a uniform tissue of approximately unit density. Some have treated separate regions of the kidney (as will be discussed in the following text) when there is nonuniform uptake, but the majority of internal dose calculations treat the whole kidney as a single region, also of approximately unit density. The skeleton is a complex mixture of bone and marrow, supported by ligaments, tendons, muscles, and cartilage. Bone has a high density (around 2.2 g/ml) and atomic number (~14) relative to soft tissue (~1 g/ml and ~7, respectively). Two types of bone are identified—trabecular and cortical. Cortical bone, also called compact bone, is a hard, dense material, while trabecular bone (also called spongy or cancellous bone) is a porous material, found in the interior of *flat* bones (e.g., ribs and skull) and in the ends of long bones such as the femur (Figure 4.1). Trabecular bone is a complicated mixture of fine bone spicules

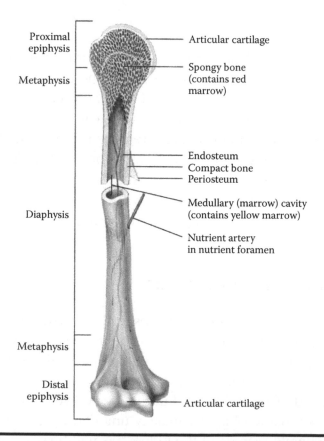

**Figure 4.1    Image of a long bone, showing several structural features.**

and tissue cavities filled with marrow. The tissues of interest to dosimetry are the *red marrow* and *endosteal cells on bone surfaces*. The dimensions of the marrow cavities vary from a few hundred micrometers to around a millimeter (Beddoe 1976). Photons will deposit energy in the skeletal tissues via secondary electrons created in interactions with atoms; electrons and alpha particles will ionize and excite atoms in the tissues. Many radioactive materials will deposit in bone because of their similarity to calcium or phosphate compounds (the principal components of bone). Some radiopharmaceuticals may be taken up in the marrow via the circulatory system. The range of alpha particles in soft tissue is of the order of micrometers, and that of electrons of moderate energy is of the order of millimeters to centimeters. Depending on whether the emitter is in the bone or marrow, an electron will cross many marrow cavities and trabeculae before depositing all of its energy. Alpha emitters on bone surfaces will irradiate cells in the marrow cavities.

Definition of the absorbed fractions is therefore not straightforward. The earliest efforts to characterize the absorbed fractions for calculation of doses to marrow and endosteal cells were proposed by Spiers and colleagues at the University of Leeds (Spiers 1949, 1966; Darley 1968, 1972; Whitwell 1973; Beddoe, Darley, and Spiers 1976b; Whitwell and Spiers 1976). They did not use geometrical models and explicit simulation of particle transport, but, in an ingenious approach, they used measurements of trabeculae and marrow cavity dimensions measured from bones taken at autopsy, using an optical scanner. They developed distributions of the chord lengths in seven bone types (cervical vertebra, lumbar vertebra, femur head, femur neck, iliac crest, parietal bone, and rib). They then used a Monte Carlo process to sample the chord lengths randomly, and estimated electron energy deposition assuming a *continuous slowing down* approximation. They provided dose factors for sources in bone volume or on bone surfaces with endosteal cells and red marrow as target regions. These factors were used by the Medical Internal Radiation Dose (MIRD)

Committee to provide dose conversion factors (called *S factors*) for 117 radionuclides. They considered four source regions (trabecular bone, cortical bone, red marrow, and yellow marrow) and four target regions (skeletal bone, red marrow, yellow marrow, and skeletal endosteum). For source/target regions not treated by the Leeds group, they applied various modeling assumptions to fill in the gap.

The International Commission on Radiological Protection (ICRP) presented a system of radiation protection for workers in its 1979 Publication 30 (ICRP 1979). They used the work of the Leeds group, but applied very conservative simplifying assumptions. They presented absorbed fractions for beta and alpha sources in bone volume or on bone surfaces, irradiating active marrow or endosteal cells (Table 4.1). They provided dose factors for hundreds of radionuclides, as well as recommended limits on radionuclide intakes, by inhalation or ingestion, based on the then-current dose limits for workers (the ICRP has no regulatory authority, but proposes dose limits periodically, which are generally followed by regulatory bodies worldwide).

In 1996, the first software program for performing internal dose calculations for radiopharmaceuticals was released. It was called MIRDOSE 2 (a version 1 preceded it, but was never released; Watson and Stabin 1984). MIRDOSE 2 used the absorbed fractions proposed in ICRP Publication 30, as they were the most recently released values, and a prominent member of the ICRP suggested their use. As they were applied in clinical nuclear medicine applications involving therapy, the highly conservative assumptions of the ICRP model were quickly seen to be suggesting very high marrow doses, when no clinical effects were being seen in patients. Keith Eckerman of Oak Ridge National Laboratory had been working on a reevaluation of the work of Spiers at this time (Eckerman 1985). He derived absorbed fractions for electron energy deposition in marrow cavities, endosteal cells, and trabeculae, for electrons of fixed starting energy (which can be

**Table 4.1   Absorbed Fractions Given in ICRP Publication 30**

| Source Organ | Target Organ | α Emitter Uniform in Volume | α Emitter on Bone Surfaces | β Emitter Uniform in Volume | β Emitter on Bone Surfaces (E ≥ 0.2 MeV) | β Emitter on Bone Surfaces (E < 0.2 MeV) |
|---|---|---|---|---|---|---|
| Trabecular bone | Bone surfaces | 0.025 | 0.25 | 0.025 | 0.025 | 0.25 |
| Cortical bone | Bone surfaces | 0.01 | 0.25 | 0.015 | 0.015 | 0.25 |
| Trabecular bone | Bone marrow | 0.05 | 0.5 | 0.35 | 0.5 | 0.5 |
| Cortical bone | Bone marrow | 0 | 0 | 0 | 0 | 0 |

generalized to any radionuclide via interpolation). The target endosteal cells were assumed to lie in a 10 μm lining of soft tissue on the surfaces of the bone regions surrounding marrow cavities. His findings mostly confirmed the work of Spiers and Whitwell and extended them. These absorbed fractions were implemented in MIRDOSE3 (Stabin 1996), and the complete model description was published by Eckerman and Stabin a few years later (Eckerman and Stabin 2000).

Various researchers at the University of Florida have done extensive work to further investigate bone and marrow dosimetry. Bouchet, Jokisch, and Bolch (1999) used the chord length distributions of Leeds group, but sampled them to perform a 3D transport simulation (the work at Leeds, updated by Eckerman, used 1D modeling approaches). This 3D simulation allowed for possible differences in particle path length and the particle's linear displacement through the media, for backscattering and for the production of delta rays and bremsstrahlung. Absorbed fractions, again for electrons of fixed starting energies, were calculated for the seven bone types from the Leeds group, with assumed trabecular bone and volume sources and dose to the marrow and a 10-μm layer of endosteal cells on bone surfaces. Some differences between these results and those of Eckerman are listed in the following text:

1. Improvements in transport at low electron energies.
2. Eckerman calculated absorbed fractions for marrow self-irradiation by multiplying the calculated absorbed fraction by the marrow cellularity in that bone, while Bouchet did not account for marrow cellularity.
3. For bone surface sources, Eckerman assumed a 2D planar source, while Bouchet assumed a 3D distribution throughout the 10-μm layer of endosteal cells.
4. The 1D Eckerman model assumes that electrons enter the endosteal layer with a uniform distribution of possible angles, while the Bouchet model assumes a cosine distribution.

The Eckerman model provides acceptable absorbed fractions for marrow self-irradiation above 200 keV, but the Bouchet model overestimated absorbed fractions in that energy region. However, the Bouchet model is more reliable at energies below about 20 keV. Working together, the groups derived a *consensus* model, combining the results of the two models and estimating values that bridge the energy gap between 20 and 200 keV (Figure 4.2) (Stabin et al. 2002).

This combined model was implemented in the OLINDA/EXM computer code (Stabin, Sparks, and Crowe 2005), the successor to the MIRDOSE series (the name was changed at the request of the MIRD Committee, to ensure that users did not mistakenly believe that it was a product from the MIRD Committee). The University of Florida group continues investigation of absorbed fractions for bone and marrow sources, using magnetic resonance imaging (MRI) of bones taken from cadavers and 3D Monte Carlo modeling methods—for example, Jokisch et al. (1998), Bolch et al. (2002), and Patton et al. (2002). Despite a significant number of publications, however, no comprehensive model that may be applied to

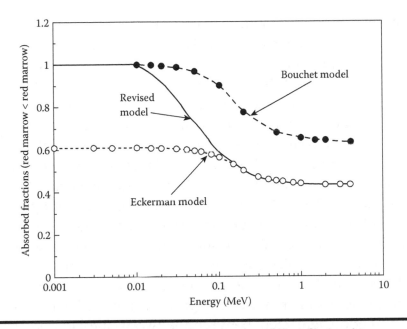

**Figure 4.2    Absorbed fractions for marrow self-irradiation in Eckerman, Bouchet, and revised (*consensus*) models.**

human phantoms of both genders or any age group has been proposed, and the OLINDA/EXM code continues to use the *consensus* model from Stabin et al. (2002).

# Models for Tumor Dosimetry

The human body models have between 20 and 40 defined organ regions, plus bone/marrow and skin. They do not explicitly define regions for tumors, as of course these can be of any size and location in the body. Several attempts have been made to define self-irradiation absorbed fractions for unit density spheres of various sizes, for photon and electron sources. These spherical constructs can be applied to any object reasonably approximated as a spherical object, but the most obvious application in nuclear medicine is to tumor dosimetry. The dose that the sphere, as a tumor, may give to or receive from other organs cannot be estimated; this is probably not a significant problem as these components are most likely small. Members of the MIRD Committee provided self-irradiation absorbed fractions for photon sources (Brownell, Ellett, and Reddy 1968; Ellett and Humes 1972) in spheres of various sizes. Siegel and Stabin (1994) provided self-irradiation absorbed fractions for electrons and beta particles for sources of fixed starting energies. The most comprehensive evaluation was given in the year 2000 by Stabin and Konijnenberg (2000). They used two different Monte Carlo codes, MCNP4B (Briesmeister 1993) and EGS4 (Bielajew and Rogers 1987), and calculated electron and photon self-irradiation absorbed fractions across a broad range of particle starting energies. Their results generally agreed very well with the results of Siegel and Stabin for electrons, and agreed well with the older MIRD results, but had a more consistent behavior across the range of energies considered. Their results showed good agreement between the two well-established and tested Monte Carlo codes, and they averaged the two values without any weighting. Two example plots, one from MCNP and one from EGS4, are shown in Figures 4.3 and 4.4.

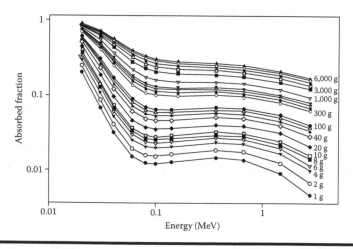

**Figure 4.3    Absorbed fractions of Stabin and Konijnenberg as function of energy for photons, EGS4. (From Stabin MG and Konijnenberg MW, 2000, *J Nucl Med*, 41, 149–160. Reproduced with permission.)**

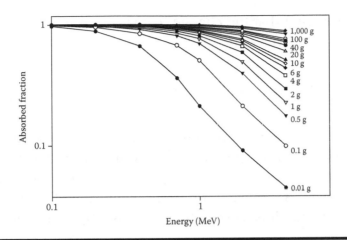

**Figure 4.4 Absorbed fractions of Stabin and Konijnenberg as function of energy for electrons, MCNP. (From Stabin MG and Konijnenberg MW, 2000, *J Nucl Med*, 41, 149–160. Reproduced with permission.)**

# Organs with Changing Organ Mass

In our derivation of Equation 1.3 in Chapter 1, we noted that, in most dosimetry applications, the only variable in our general equation that changes with time is activity. In doing dosimetry for human populations from radionuclide releases to the environment, doses may need to be integrated over ages, during which organs' mass and shape may change. This is a challenging calculation! In nuclear medicine, however, doses are delivered over times that are short compared to any times over which a person's size and shape will vary. In two applications, however, organ size may change during the delivery of the radiation dose: (1) radioiodine treatment of hyperthyroidism, in which the goal is to shrink the thyroid, and (2) radiation therapy of tumors. Traino et al. (2000) treated the mathematics for the first case. They gave expressions for the general case, in which mass may change during iodine uptake and clearance, and the case in which thyroid mass changes only during the clearance phase.

They proposed that the change in mass as a function of time, $m(t)$, may be given by:

$$m(t) = \left[ 2\left( \frac{k\, A_m}{c} \exp(-c(t-T)) - \frac{k\, A_m}{c} + 0.5 * (m_T)^2 \right) \right]^{0.5} \quad (4.1)$$

Here, $A_m$ is the maximum thyroid uptake, $c$ is the rate constant characterizing clearance of iodine from the thyroid, $k$ is a rate constant characterizing the reduction of thyroid mass, and $m_T$ is the mass of the thyroid at time $T$ after the therapy administration. The total dose $D_T$ is then given as:

$$D_T = \sigma A_0 \left( \frac{a}{m_0} \int_0^T t\, dt + b \int_T^\infty \frac{\exp(-ct)}{m(t)} dt \right) \quad (4.2)$$

In this model, $a$ and $b$ are patient-specific parameters related to the thyroid uptake and clearance phases.

## Special Body Kinetic Models

In most cases, our evaluation of the kinetics of activity uptake and/or clearance from individual organs or the total body is established from data extrapolated from animals or measured in patients using external imaging devices. In two cases, we use a mathematical model to estimate the integrated activity in certain organs, for reasons that will be explained.

### The Voiding Urinary Bladder

When activity is excreted from the body in the urine, the function that describes it usually consists of one or more exponential terms. Fitting observed activity levels in the urinary bladder is not helpful, as the bladder fills and empties repeatedly, and measurements are too infrequently gathered to characterize this time–activity curve. Material leaving the body is most often governed by first-order processes, which mean that the

retention (in the body) can be expressed as a function such as
$A^*\exp(-\lambda\ t)$. Therefore, the time–activity curve for the bladder
takes the form of $A^*(1 - \exp(-\lambda\ t))$, but the curve is periodically
interrupted by voiding and goes to zero (or nearly zero) and
then begins to accumulate again, as shown in Figure 4.5.

What is needed is a characterization of the values $A$ and
$\lambda$ (in real situations, there may be more than one term in the
equation, but let us just consider one for now). In a particularly
ingenious derivation, Walt Snyder and colleagues showed that
the number of disintegrations occurring in the bladder could be
given in such cases by a single equation:

$$N = A_0 \sum_i f_i \left[ \frac{1 - e^{(-\lambda_i T)}}{(\lambda_i)} - \frac{1 - e^{-(\lambda_i + \lambda_p)T}}{\lambda_i + \lambda_p} \right] \left[ \frac{1}{1 - e^{-(\lambda_i + \lambda_p)T}} \right] \quad (4.3)$$

Here, $A_0$ is the initial activity entering the body, $\lambda_p$ is the
physical decay constant of the radionuclide, $\lambda_i$ is the biological
removal constant for the fraction of activity $f_i$ leaving the body
via the urinary pathway, and $T$ is the bladder-voiding interval,

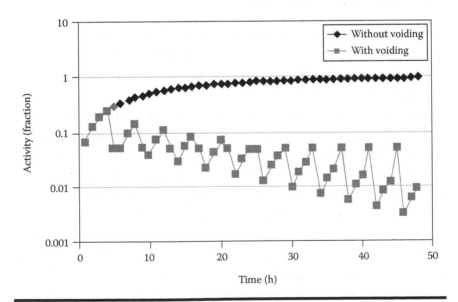

**Figure 4.5  Urinary bladder time–activity curve.**

assumed to be constant. If we have all the activity in the body passing out through the urinary pathway with a 1-h half-time, for example, our $f$ would be 1.0, and $\lambda_b$ would be $0.693/1$ h $= 0.693$ h$^{-1}$. Let us say we have 40% passing out through the GI tract, and 60% through the urinary pathway, with two-thirds of the urinary clearance having a biological half-time of 1 h, and one-third with a biological half-time of 10 h. Then, $f_1$ would be 0.4 and $\lambda_{b1}$ would be 0.693 h$^{-1}$, and $f_2$ would be 0.2 and $\lambda_{b2}$ would be 0.0693 h$^{-1}$. These parameters are not particularly hard to derive; one must either measure the total body retention or the cumulative urinary excretion and fit a function, either of the form $A*\exp(-\lambda\,t)$ (in the former case) or $A*(1-\exp(-\lambda\,t))$ (in the latter case). Again, the equation may have more than one term, depending on the data observed. If there is GI excretion, this complicates the use of whole-body retention data, unless intestinal activity is somehow excluded from the images. However, in either case, the complication can be overcome by careful data gathering and inspection of the results.

## The Gastrointestinal Tract

Ingestion of radiopharmaceuticals to study gastrointestinal tract function is less common than injection, but is done regularly. More commonly, radiopharmaceuticals may be taken up in the liver and excreted into the intestines via the biliary pathway. The stomach and intestines are difficult to image, due to their complex geometry. A standardized kinetic model of the GI tract was first proposed by Dolphin and Eve (1966) and was adopted by the International Commission on Radiological Protection in ICRP Publication 30 (ICRP 1979). Four sections of the GI tract were defined (Figure 4.6), having separate kinetics, with activity in the contents passing through with standard rate constants. The walls of the various sections are the important target tissues. A more detailed and realistic model has been recommended recently by the ICRP, named the human alimentary tract (HAT) model (Figure 4.7).

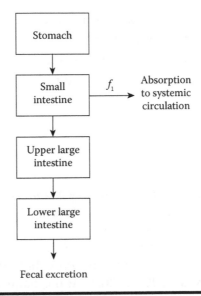

**Figure 4.6    ICRP 30 gastrointestinal tract model.**

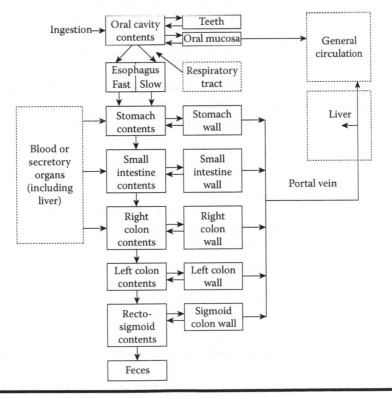

**Figure 4.7    ICRP human alimentary tract model.**

This model has more compartments, includes some non-first-order kinetic components, models age-dependent compartment transfer rates, treats liquid and solid materials differently, and calculates doses to segments not treated in the ICRP 30 model.

# Bibliography

Beddoe AH. 1976. *The Microstructure of Mammalian Bone in Relation to the Dosimetry of Bone-Seeking Radionuclides.* University of Leeds, Leeds, UK.

Beddoe AH, Darley PJ, and Spiers FW. 1976. Measurements of trabecular bone structure in man. *Phys Med Biol*, 21, 589–607.

Bielajew A and Rogers D. 1987. PRESTA: the parameter reduced electron-step transport algorithm for electron Monte Carlo transport. *Nucl Instrum Methods*, B18, 165–181.

Bolch WE, Patton PW, Rajon DA, Shah AP, Jokisch DW, and Inglis BA. 2002. Considerations of marrow cellularity in 3-dimensional dosimetric models of the trabecular skeleton. *J Nucl Med*, 43(1), 97–108.

Bouchet LG, Jokisch DW, and Bolch WE. 1999. A three-dimensional transport model for determining absorbed fractions of energy for electrons within trabecular bone. *J Nucl Med*, 40, 1947–1966.

Briesmeister J. 1993. *MCNP: A General Monte Carlo N-Particle Transport Code—MCNP Users Manual.* Los Alamos National Laboratory, Los Alamos, NM.

Brownell G, Ellett W, and Reddy R. 1968. *Absorbed Fractions for Photon Dosimetry.* MIRD Pamphlet No. 3. Society of Nuclear Medicine New York, NY.

Darley PJ. 1968. Measurement of linear path length distributions in bone and bone marrow using a scanning technique. *Proceedings of the Symposium on Microdosimetry* (pp. 509–526). EAEC Report EUR 3747 Ispra, Italy.

Darley PJ. 1972. *An Investigation of the Structure of Trabecular Bone in Relation to the Radiation Dosimetry of Bone-Seeking Radionuclides.* University of Leeds, Leeds, UK.

Dolphin GW and Eve IS. 1966. Dosimetry of the gastrointestinal tract. *Health Phys*, 12(2), 163–72.

Eckerman KF. 1985. Aspects of the dosimetry of radionuclides within the skeleton with particular emphasis on the active marrow. In *Proceedings of the Fourth International Radiopharmaceutical Dosimetry Symposium*, edited by AT Schlafke-Stelson and EE Watson (pp. 514–534). ORAU, Oak Ridge, TN.

Eckerman KF and Stabin MG. 2000. Electron absorbed fractions and dose conversion factors for marrow and bone by skeletal regions. *Health Phys*, 78, 199–214.

Ellett W and Humes R. 1972. *Absorbed Fractions for Small Volumes Containing Photon-Emitting Radioactivity*. Society of Nuclear Medicine. MIRD Pamphlet No. 8. New York, NY.

ICRP (International Commission on Radiological Protection). 1979. Limits for intakes of radionuclides by workers Publication 30, Part 1 International Commission on Radiological Protection. Pergamon Press, Oxford.

Jokisch DW, Patton PW, Inglis BA, Bouchet LG, Rajon DA, Rifkin J, and Bolch WE. 1998. NMR microscopy of trabecular bone and its role in skeletal dosimetry. *Health Phys*, 75, 584–596.

Patton PW, Rajon DA, Shah AP, Jokisch DW, Inglis BA, and Bolch WE. 2002. Site-specific variability in trabecular bone dosimetry: considerations of energy loss to cortical bone. *Med Phys*, 29(1), 6–14.

Siegel JA and Stabin MG. 1994. Absorbed fractions for electrons and beta particles in spheres of various sizes. *J Nucl Med*, 35, 152–156.

Spiers FW. 1949. The influence of energy absorption and electron range on dosage in irradiated bone. *British J of Radiol*, 22, 521–533.

Spiers FW. 1966. A review of the theoretical and experimental methods of determining radiation dose in bone. *British J of Radiol*, 39, 216–221.

Stabin MG. 1996. MIRDOSE: personal computer software for internal dose assessment in nuclear medicine. *J Nucl Med*, 37, 538–546.

Stabin MG and Konijnenberg MW. 2000. Re-evaluation of absorbed fractions for photons and electrons in spheres of various sizes. *J Nucl Med*, 41, 149–160.

Stabin MG, Eckerman KF, Bolch WE, Bouchet LG, and Patton PW. 2002. Evolution and status of bone and marrow dose models. *Cancer Biother Radiopharm*, 17(4), 427–434.

Stabin MG, Sparks RB, and Crowe E. 2005. OLINDA/EXM: the second-generation personal computer software for internal dose assessment in nuclear medicine. *J Nucl Med*, 46, 1023–1027.

Traino AC, Martino FD, Lazzeri M, and Stabin MG. 2000. Influence of thyroid volume reduction on calculated dose in radioiodine therapy of Graves' hyperthyroidism. *Phys Med Biol*, 45, 121–129.

Watson EW and Stabin MG. 1984 (Feb 5–9). BASIC alternative software package for internal dose calculations. In *Computer Applications in Health Physics, Proceedings of the 17th Midyear Topical Symposium of the Health Physics Society* (pp. 7.79–7.86). Pasco, WA.

Whitwell JR. 1973. *Theoretical Investigations of Energy Loss by Ionizing Particles in Bone.* University of Leeds, Leeds, UK.

Whitwell JR and Spiers FW. 1976. Calculated beta-ray dose factors for trabecular bone. *Phys Med Biol*, 21, 16–38.

*Chapter 5*

# Dose Calculations for Diagnostic Pharmaceuticals

The safety of consumer products is overseen by various government agencies. In the United States, the US Food and Drug Administration (FDA) sets standards for the use of lasers (21CFR) and other non-ionizing radiations, food irradiation, and pharmaceuticals. In the European Union, a two-step process is followed—a clinical trial application and a marketing authorization application. Clinical trial applications are carried out at the member state level, while marketing authorization applications are approved at both the member state level and in centralized authorities (Vishal et al. 2014). There are many safety concerns to be addressed in the approval of a radio-pharmaceutical, for either diagnostic or therapeutic applications; here, we will just discuss issues related to the radiation doses that may be received.

# United States Regulatory Process

Approval of a new medical imaging agent includes several phases (from the USFDA web site):

■ *A preclinical phase*, in which studies in an appropriate animal species are carefully planned and executed, to provide a preliminary assessment of the possible radiation doses expected in human subjects

■ *Phase 1* studies of medical imaging agents, which are designed to obtain pharmacokinetic and human safety assessments, based on a single mass administration and escalating mass administrations of the drug or biological product.

■ *Phase 2* studies of medical imaging agents, which include:
  - Refining the agent's clinically useful mass dose and radiation dose ranges or dosage regimen (e.g., bolus administration or infusion) in preparation for phase 3 studies
  - Answering outstanding pharmacokinetics and pharmacodynamics questions
  - Providing preliminary evidence of efficacy and expanding the safety database
  - Optimizing the techniques and timing of image acquisition
  - Developing methods and criteria by which images will be evaluated
  - Evaluating other critical questions about the medical imaging agent

■ *Phase 3* studies are designed to (1) confirm the principal hypotheses developed in earlier studies, demonstrating the efficacy of the compound and the method employed; (2) verify the safety of the use of the medical imaging agent; and (3) validate the necessary instructions for use of the compound and for imaging in the population for which the agent is intended.

In Chapter 3, we outlined the steps and calculations for the use of animal models to try to predict human radiation dose estimates. We noted the paper by Sparks and Aydogan (1999) which found that data collected in preclinical trials are particularly poor at predicting human dosimetry. Nonetheless, this is a necessary step in the approval process for any new radiopharmaceutical. The Phase 1, 2, and 3 clinical trials are very expensive, and animal studies make up a small percentage of the overall approval process (Figure 5.1). They are, nonetheless, expensive and time consuming for the developer.

Clinical trials with human subjects (healthy volunteers or patients—therapeutic agents may only be tested in patients with the disease to be treated) involve *dose escalation* studies—starting with a small amount of activity (*microdosing*) and increasing the

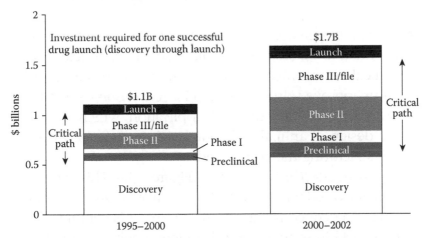

The figure shows one estimate of the total investment required to "launch" (i.e., market) a successful drug in two time periods. Most of the recent cost increases are within the "critical path" development phase, between discovery and launch.

The overall increase between 1995–2000 and 2000–2002 is estimated to be 55 percent.

**Figure 5.1** **Reported costs of new drug approval, circa 2000. (From Windhover's *In vivo*: The Business & Medicine Report, Bain drug economics model, Windhover Information Inc, Norwalk, CT, 2003. Available at http://www.bain.com/bainweb/PDFs/cms/Marketing/ rebuilding_big_pharma.pdf)**

dose of the compound (here *dose* indicates the amount of the drug, not radiation dose) to evaluate possible toxicity. For therapeutic agents, radiotoxicity from the radiation dose delivered to sensitive normal tissues is also of concern. For many agents—for example, radiolabeled monoclonal antibodies—the red marrow is the organ most likely to be affected. Monitoring of blood elements such as platelets and white blood cells is performed to assess toxicity (Wiseman et al. 2003). For radiolabeled peptides, marrow dose may be important, but renal toxicity may also be observed; this may be monitored by evaluating patients' changes in creatinine clearance rate (Bodei et al. 2011).

## Types of Approvals

Medical imaging agents are submitted for approval in:

- *Investigational new drug (IND) applications*—These are intended to be studies of limited scope, conducted in clinical environments, to demonstrate the properties of an unapproved drug (Figure 5.2). A *sponsor*—some firm or institution—files the IND application, which describes the design of animal or human trials and the process for manufacturing the compound.
- *New drug applications (NDAs)* (Figure 5.3)—This is the full process for bringing a new drug to market, with the four phases described in the preceding text. It is common for this to be held in several medical centers simultaneously. Having all the institutions follow the study protocol in the same way can be quite challenging. These trials take many years and can cost several billion US dollars (VanBrocklin 2008).
- *Biologics license applications (BLAs)*—This is the formal drug approval process for a radiopharmaceutical that has *biologic* properties—for example, a monoclonal antibody.
- *Abbreviated NDAs (ANDAs)*—These can be used for the approval of generic drugs, those that are similar to other previously approved drugs.

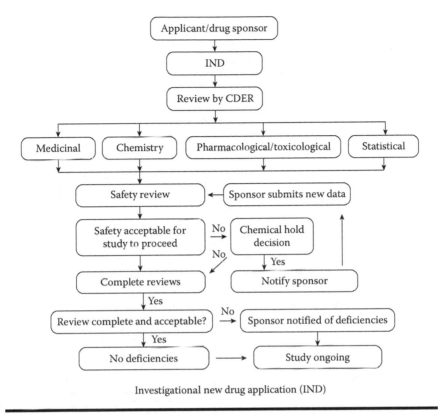

Investigational new drug application (IND)

**Figure 5.2  Flowchart for an investigational new drug. (From Vishal P et al., 2014, *Int J Drug Regulatory Affairs*, 2, 1–11.)**

# European Regulatory Process

As noted in the preceding text, in Europe, there are two processes—clinical trial applications (CTAs), which are carried out at the member state level, and marketing authorization applications (MAAs), which are approved at both the member state level and in centralized authorities (Vishal et al. 2014). A CTA is roughly equivalent to an IND in the United States. This submission may be made country by country, or using a Voluntary Harmonisation Procedure (VHP). Starting on June 16, 2014, new clinical trial regulation entered into force will be implemented no later than May 28, 2016. This was an important step in the harmonization of the CTA process in the EU.

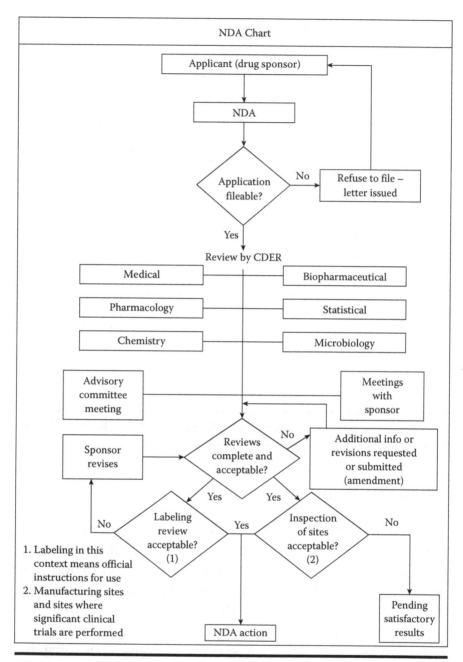

**Figure 5.3   Flowchart for a new drug approval. (From Vishal P et al., 2014, *Int J Drug Regulatory Affairs*, 2, 1–11.)**

After this date, a single electronic application will be submitted via an EU portal. The sponsor will be required to propose a Reporting Member State (RMS), who will conduct the assessment in coordination with the other Concerned Member States (CMS).

MAAs may be submitted either by a centralized procedure (CP), decentralized procedure (DCP), or purely national procedure (NP). If a centralized procedure is used, the application is reviewed by the European Medicines Agency (EMA), and the marketing authorization is valid in the entire European Union (EU). This is required for products involving biotechnology (c.g., genetic engineering) and for pharmaceuticals intended to treat cancer, HIV/AIDS, diabetes, neurodegenerative disorders or autoimmune diseases, and other immune disorders. Decentralized procedures can be used for approval of other drugs in one or more EU countries. A process called mutual recognition process (MRP) allows applicants to obtain marketing authorizations in CMSs other than the *Reference Member State* where approval has previously been received (Figures 5.4 and 5.5).

The remaining products may be approved via other types of procedures, in which the application can be submitted to one or more EU countries, depending on the marketing interest of the company. In both MRP and DCP, an RMS is assigned. The RMS leads the procedure and performs the assessment of the application, while CMSs adopt the RMS's opinion. Then, national marketing authorizations are granted in each of the countries separately.

# Study Design

## *Number of Subjects and Time Points*

Considering the expense of performing trials to obtain dosimetry data, it is important that the study design is carefully

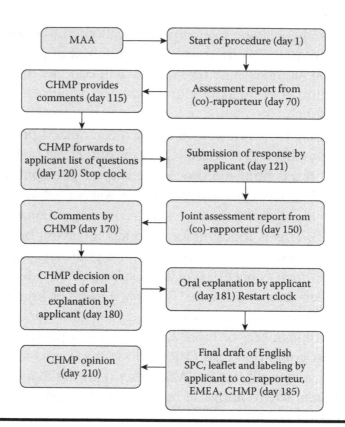

**Figure 5.4 Flowchart for a centralized procedure. (From Vishal P et al., 2014, *Int J Drug Regulatory Affairs*, 2, 1–11.)**

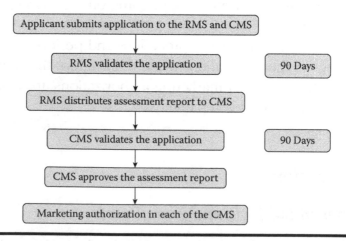

**Figure 5.5 Flowchart for a mutual recognition procedure. (From Vishal P et al., 2014, *Int J Drug Regulatory Affairs*, 2, 1–11.)**

thought out in advance. The number of *subjects*, either animals or humans, is difficult to establish absolutely. Animals are generally sacrificed at different times after administration of a radiopharmaceutical. A minimum of three animals per time point is desirable to obtain any kind of reasonable statistics. Animals can be imaged; this is more common in primates or other larger species than in mice or rats. The problem with using image data is that it is generally necessary to administer anesthetic to prevent the animal from moving. Anesthetics can alter the biodistribution and excretion of the radiopharmaceuticals. The number of human subjects has to be reasonable— probably a minimum of 10 subjects, as the variability of human metabolism is considerable. Figures 5.6 and 5.7 show

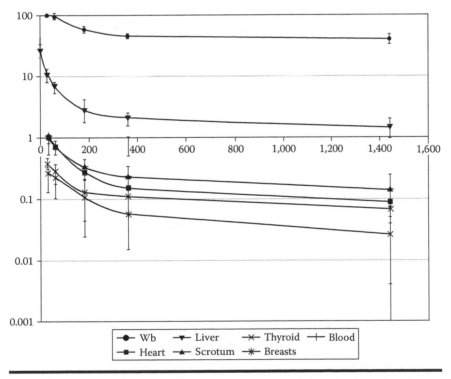

**Figure 5.6**  Uptake and retention of ⁹⁹ᵐTc-RP527, a gastrin-releasing peptide (GRP) agonist for the visualization of GRP receptor expressing malignancies in various subjects. (From Van de Wiele C et al., 2001, *J Nucl Med*, 42, 1722–1727.)

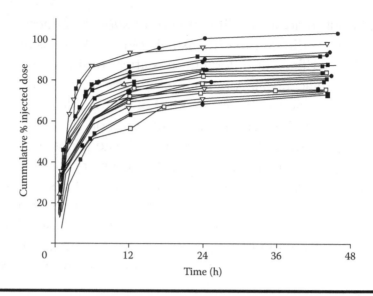

**Figure 5.7  Cumulative excretion of [166]Ho-DOTMP in 12 subjects (six female, six male) with multiple myeloma. (From Breitz HB et al., 2006, *J Nucl Med*, 47, 534–542.)**

observed variability in organ uptake and retention of a gastrin-releasing peptide (Van de Wiele et al. 2001) and urinary excretion of [166]Ho-DOTMP (Breitz et al. 2006). In the case of [90]Y-Zevalin, data were gathered on 179 patients (Wiseman et al. 2003); this was an unusually large patient cohort!

Choosing the number and spacing of time points to assess the biokinetics of a radiopharmaceutical is another essential aspect of the clinical study design. An overall limitation is the physical half-life of the radionuclide. Fluorine-18 is a very popular imaging agent; its 110 min half-life limits the possible times for imaging to perhaps 12 h. Siegel et al. (1999) wrote:

> To determine the activity–time profile of the radioactivity in source regions, four questions need to be answered:
>
> ■ What regions are source regions?
> ■ How fast does the radioactivity accumulate in these source regions?

- How long does the activity remain in the source regions?
- How much activity is in the source regions?

The first question concerns identification of the source regions, while the second and third questions relate to the appropriate number of measurements to be made in the source regions as well as the timing of these measurements. The fourth question is addressed through quantitative external counting and/or sampling of tissues and excreta. Each source region must be identified and its uptake and retention of activity as a function of time must be determined. This provides the data required to calculate cumulated activity or residence time in all source regions. Each region exhibiting significant radionuclide uptake should be evaluated directly where possible. The remainder of the body (total body minus the source regions) must usually be considered as a potential source as well. Mathematical models that describe the kinetic processes of a particular agent may be used to predict its behavior in regions where direct measurements are not possible, but where sufficient independent knowledge about the physiology of the region is available to specify its interrelationship with the regions or tissues whose uptake and retention can be measured directly. The statistical foundation of a data acquisition protocol designed for dosimetry requires that an adequate number of data points are obtained and that the timing of these points be carefully selected. As the number of measurements increases, the confidence in the fit to the data and in the estimates of unknown parameters in the model is improved. As a heuristic or general rule of thumb, at least as many data points should be obtained as the number of initially unknown variables in the mathematical curve-fitting function(s) or in the compartmental model applied to the data set. For example, each exponential term in a multi-exponential curve-fitting function requires two data points to be adequately characterized. On the other hand, if it is known a priori that the activity retention in a region can be accurately represented by a monoexponential function, restrictions on sampling times are less stringent as long as enough data points are obtained to derive the fitted function. Because of problems inherent in the collection of patient data (e.g., patient motion, loss of specimen, etc.), the collection of data above the necessary minimum is advisable.

The kinetics of some radionuclides (e.g., $^{131}$I-Bexxar) may be characterized by a single exponential function; more commonly, organ, tumor, and total body kinetics need two or more exponential functions to describe the retention and excretion of a compound. In human imaging studies, it is desirable to obtain an image as soon after injection as possible, before any excretion occurs and negligible radioactive decay has occurred, to establish the level of counts that represents 100% of the injected activity. This is particularly important in planar or SPECT imaging, as the procedures needed to establish an absolute calibration of counts to activity are seldom followed in many medical centers. PET imaging is generally quantitative, but usually only a partial body image is collected, from the head to the upper portion of the legs, so 100% of the counts are not in the images. So, a general method is to get one or two early time points, and then as many time points as possible to characterize later phases of the kinetics, over several physical half-lives of the radionuclide. The desire for abundant information must always be balanced against the costs (of the animals or imaging studies), inconvenience to patients and staff, normal operating hours for the clinic, and other considerations. Siegel et al. (1999) noted four typical kinetic models that treat biokinetic data:

- Instantaneous uptake (wash-in) with no biologic removal
- Instantaneous uptake with removal by both physical decay and biologic elimination (washout)
- Non-instantaneous uptake with no biologic removal
- Non-instantaneous uptake with removal by both physical decay and biologic elimination

They also provided quantitative data on the errors that will occur if an inadequate number of data points are obtained (Figure 5.8).

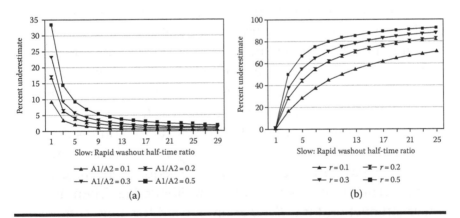

**Figure 5.8  Percent error in organ time–activity integrals when an early, rapid washout phase is neglected (a), or when a long-term retention phase is neglected (b). Values of *A* are the fractional activity in different retention components, and *r* is the activity removed from an organ to activity in the total body. (From Siegel J et al., 1999, *J Nucl Med*, 40, 37S–61S.)**

# Imaging Methods

## Planar Imaging with Scintillation Cameras

The simplest method of quantitative imaging for dosimetry involves the use of a dual-headed scintillation camera to obtain 180° opposed planar images, usually of the entire patient, from head to toe. The posterior image can be inverted using image management software, so that organs are aligned with the anterior image. This is called the *conjugate view* imaging procedure. Regions of interest (ROIs) may be manually drawn over identifiable organ regions, tumors, and the whole body. The activity in a given region is calculated as:

$$A_j = \sqrt{\frac{I_A I_P}{exp\left(-\mu_e t\right)}}\, \frac{f_j}{C} \qquad (5.1)$$

where $I_A$ and $I_P$ are the observed counts over a given period for a given region of interest (ROI) in the anterior and posterior projections (counts/time); $t$ is the patient thickness over the ROI; $\mu_e$ is the effective linear attenuation coefficient for the radionuclide, camera, and collimator; $C$ is the system calibration factor (counts/time per unit activity); and the factor $f$ represents a correction for the source region attenuation coefficient ($\mu_j$) and source thickness ($t_j$) (i.e., source self-attenuation correction). The linear attenuation coefficient for the radionuclide of interest, $\mu_e$, may be established by placing increasing thicknesses of tissue-equivalent materials over a small source containing the radionuclide. Then, the linear attenuation for a *flood* source, used for daily camera uniformity tests, often a $^{57}$Co source placed under the imaging table, is established similarly. The effective thickness of the patient is then established by imaging the $^{57}$Co source with and without the patient on the table. The total patient thickness, $t$, over the drawn ROIs may be used, regardless of the depth of the organ imaged, because whatever thickness of tissue $t_1$ is between the organ and one camera head and thickness $t_2$ is between the organ and the other camera head (Figure 5.9), and

$$exp(-\mu_e t_1) \times exp(-\mu_e t_2) = exp(-\mu_e t) \qquad (5.2)$$

The preceding equation involves an attempt to correct the counts in the image for attenuation. Other corrections that should be made include:

- *Background subtraction*—Organs in the body that have significant uptakes of the radiopharmaceutical will be identified as *source* regions and assigned ROIs. An ROI for the total body will also be created. Inside the body, there will be a somewhat uniform level of *background* activity—in blood, extracellular fluids, and other struc-tures. A *background* ROI or ROIs should be drawn to subtract activity in these overlying or underlying tis-sues from the counts extracted for a given ROI. There is no exact method for drawing these background ROIs. Pereira et al. (2010) experimented with a number of strategies for background ROI, drawing around struc-tures in a water-filled phantom with known amounts of activity. Their best results were found with small regions partially surrounding the structures (Figure 5.10). Outside of the body, there will be very low levels of scattered photons; activity in *total body* may be corrected for this

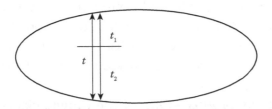

**Figure 5.9** **Tissue thicknesses of an object at an arbitrary depth in a patient.**

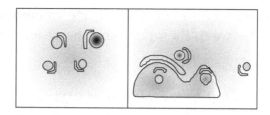

**Figure 5.10   Example ROIs and background ROIs suggested by Pereira et al. (2010) around regions filled with known amounts of activity in a physical phantom.**

background with an ROI placed outside of the body, in a region that appears typical of the average background level.

■ *Scattered photons*—The counts in a nuclear medicine image are obtained within a predefined *window* of acceptable energies. Typical values are ±10% around the photopeak energy (e.g., for the 365 keV photopeak of $^{131}$I, a 20% window would be from about 328 keV to 401 keV). Any photopeak will rest on a *baseline* of photons scattered from higher energies, that is, Compton-scattered photons that happened to fall into this energy range. Photopeaks for scintillation detectors are quite broad, so even photons that were actually from the photopeak energy may be scattered slightly to lower values within the window, and photons from higher-energy photons may as well. The most common method to correct for this is the triple-energy window technique (Siegel et al. 1999), in which small windows are placed just below and just above the photopeak, and counts from these windows are subtracted from the photopeak counts, sometimes using weighting factors to account for the fact that Compton scatter contributions generally decrease with increasing energy (Figure 5.11).

■ *Overlapping organs*—One of the drawbacks of planar imaging is that organs in the body with significant amounts of activity may partially or completely be within

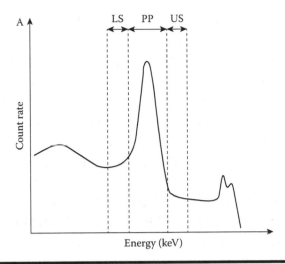

**Figure 5.11    The triple-energy window technique for correction of scattered photons in a scintillation camera photopeak. (From Siegel J et al., 1999, *J Nucl Med*, 40, 37S–61S.)**

the ROI of another organ of interest. A common example is the liver overlying part or all of the right kidney. There are no perfect methods to correct for this. In the case of kidneys, some investigators have measured activity in the left kidney, assuming it suffers no overlap issues, and just multiplied this by a factor of two. If some part of the organ can be assigned an ROI without overlap, and some estimate can be made of the fraction of the organ's total mass or volume in that ROI, the counts can be adjusted with this fraction. If red marrow has a notable uptake of the radiopharmaceutical, only a small portion of the marrow can be assigned an ROI. One may use a small ROI around one or two lumbar vertebrae, for example, and assume what fraction of total marrow is in these regions. Some have also assigned marrow activity using a factor to relate blood concentrations to marrow concentrations. If marrow binding is low, imaging methods will produce similar results to blood-based methods (Siegel, Stabin, and Brill 2002).

■ *Radioactive decay*—It is of course important to perform correction for radioactive decay in any study. If the data are presented for dosimetric analysis without decay correction, this is fine; the fitted data will represent the effective half-time for the clearance (Chapter 1). If they are corrected for decay, the half-lives obtained will be the biological half-times and can be corrected for decay during integration of the data. It can be an area of confusion when someone says *the data were decay corrected*; it is important to be clear about which case is being referred to.

■ *Deadtime*—It is not usual for diagnostic nuclides to cause any deadtime problems for nuclear medicine cameras, but a therapeutic agent may cause this phenomenon, during calibration or patient imaging. The issue is that, at high levels of activity, there are so many events being processed by the camera's electronics that real counts are being lost during the processing. If this is anticipated to be a problem, a correction can be made using the *decaying source method*; this involves counting a source that will provide deadtime at early times, but, as it decays, it will reach the point after which the camera will not suffer deadtime count losses. At these times, the source will decay exponentially, which will be plotted as a straight line on a semi-logarithmic plot. At times, when this curve is not linear, a correction can be made by extrapolating the linear portion of the curve (Figure 5.12).

## Tomographic Imaging

Obtaining quantitative data for dosimetry from tomographic methods (single-photon emission computed tomography, SPECT, or positron emission tomography, PET) requires significantly more effort than planar methods. Centers generally calibrate PET scanners to give quantitative uptake in organs and tumors, so the data are thus

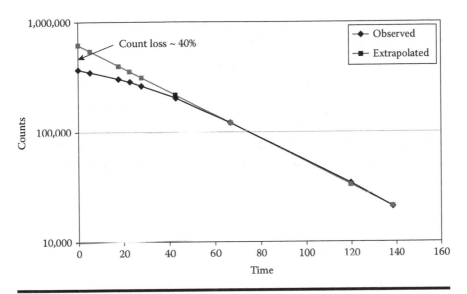

**Figure 5.12** The *decaying source method* to correct for scintillation camera deadtime.

quantitative when obtained, and one simply needs the calibration factor to convert counts to activity. However, the quantification is much more difficult, as *volumes of interest* (VOIs) are established by drawing regions on multiple *slices* of the tomographic reconstruction (Figure 5.13). Quantification from SPECT images is much more difficult.

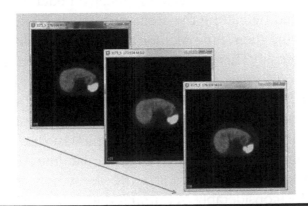

**Figure 5.13** Drawing volumes of interest on tomographic images.

Dewaraja et al. (2012) provided a thorough presentation of the procedures needed to obtain quantitative SPECT data for dosimetry calculations, including choice of collimator, choice of energy windows, reconstruction methods, attenuation and scatter correction, deadtime corrections, compensation for other image-degrading effects, choice of target regions, corrections for dose nonuniformities, and other aspects to be considered to obtain quantitative information in individual voxels used to define source and target regions.

## Kinetic Modeling

Once suitable kinetic data have been gathered, we have estimates of activity in the body, organs, and possibly tumors. In order to calculate dose estimates, we have to integrate the time–activity curves to obtain the area under the curve (AUC) for all regions of interest (Chapter 1). There are three common approaches to performing this integration.

- ▪ *Direct integration*—Making a direct calculation of the area under the curve between any two points can be performed simply, typically by assuming that the area between two points is represented by a trapezoidal area. (This author has actually received data that were measured by cutting the areas printed on paper and measuring them on a Mettler balance! The data made good sense!) This is very easy, and can be performed quickly. The major concern about this method is that it only accounts for activity during the period of observation. Some allowance has to be made for possible area under the time activity curve after the last data point. Assigning no activity may underestimate the AUC considerably. Assuming only radioactive decay may overestimate the AUC considerably. Assuming the curve goes linearly to zero with the slope of the last two points may also underestimate the AUC. In any case, the key is to state clearly what assumption was used.

■ *Least squares analysis*—As almost every time–activity curve for radiopharmaceuticals is well represented by exponential functions, probably the most common approach is to fit the data to one or more exponential terms. One must have a minimum of two data points for each phase of the kinetic curve (ingrowth or decay), as two variables must be resolved (the fraction of activity associated with the term and its half-time). Thus, during the study design phase, it is important to have some idea of how many phases of uptake or clearance to expect, and over what time frame. This is, of course, difficult with a new radiopharmaceutical; one starts with the limitation of the physical half-life of the radionuclide and then tries to gather data over several half-lives at several points that can characterize the kinetics. Most curves are well fitted with one or two exponentials. The curve will have the form of $a_1 \exp(-b_1 t) + a_2 \exp(-b_2 t)$, and the integral over infinite time is just $a_1/b_1 + a_2/b_2$.

Stabin (2008) provided an example and integrated the data by both methods given in the preceding text (Table 5.1).

**Table 5.1  Hypothetical Retention Data for an Organ**

| Time (h) | Activity (Bq) |
|----------|---------------|
| 0        | 100           |
| 0.5      | 72            |
| 1        | 35            |
| 2        | 24            |
| 4        | 20            |
| 6        | 15            |
| 10       | 12            |

Using trapezoidal integration, the AUC was 232 Bq – h; using least squares fitting, the value was 543 Bq – h, showing that a considerable amount of area is probably under the curve after the last datum.

■ *Compartmental modeling*—The most complex approach is to set up a system of compartments representing different organs of the body, linked by transfer rate coefficients (Figure 5.14). The system can be solved using software codes to solve the differential equations that represent the system. The result of this solution is a series of exponential terms that can be integrated as in the last section, to obtain the AUCs for each organ. This can be quite difficult to solve at times, and generally takes more time. The advantage is that all activity in the system is conserved at all times. Several computer codes are available for this kind of analysis, including SAAM II (University of Washington, Seattle, Washington, USA), Stella Isee Systems (Lebanon, New Hampshire, USA), PMOD (PMOD Technologies, Zurich, Switzerland), Simple (Gambhir et al. 1995), and others.

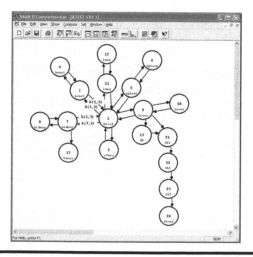

**Figure 5.14 Compartment model for a hypothetical radiopharmaceutical.**

## Dose Calculations for Patients

Generally, dose estimates based on standardized phantoms are sufficient to obtain the needed dose estimates for regulatory approval and use in the radiopharmaceutical package insert (Figure 5.15). On a day-to-day basis, once a diagnostic

---

513121-0710                                                                July 2010

**Lantheus**
**Medical Imaging**
331 Treble Cove Road
N. Billerica, Massachusetts 01862
USA

# CARDIOLITE®

**Kit for the Preparation of**
**Technetium Tc99m Sestamibi for Injection**

FOR DIAGNOSTIC USE

**HIGHLIGHTS OF PRESCRIBING INFORMATION**

**These highlights do not include all the information needed to use CARDIOLITE® safely and effectively. See full prescribing information for CARDIOLITE®.**

CARDIOLITE®, Kit for the Preparation of Technetium Tc99m Sestamibi for Injection. Initial U.S. Approval: December, 1990

----------------------------RECENT MAJOR CHANGES----------------------------

Use in specific populations (8.3)                                           10/2007

----------------------------INDICATIONS AND USAGE----------------------------

CARDIOLITE® is a myocardial perfusion agent indicated for:

• detecting coronary artery disease by localizing myocardial ischemia (reversible defects) and infarction (non-reversible defects)
• evaluating myocardial function and developing information for use in patient management decisions

----------------------------DOSAGE AND ADMINISTRATION----------------------------

• For myocardial imaging: The suggested dose range for I.V. administration of CARDIOLITE® in a single dose to be employed in the average patient (70 Kg) is 370–1110 MBq (10–30 mCi).
• For breast imaging: The recommended dose range for I.V. administration of MIRALUMA® is a single dose of 740–1110 MBq (20–30 mCi).

----------------------------DOSAGE FORMS AND STRENGTHS----------------------------

• CARDIOLITE®, Kit for the Preparation of Technetium Tc99m Sestamibi for Injection is supplied as a lyophilized mixture in a 5 mL vial.

----------------------------CONTRAINDICATIONS----------------------------

• None known

---

**Figure 5.15  Sections of a sample radiopharmaceutical package insert.** (From Cardiolite, http://www.nuclearonline.org/PI/Cardiolite.pdf)

*(Continued)*

**Table 1.0. Radiation Absorbed Doses from Tc99m Sestamibi**

Estimated Radiation Absorbed Dose

| | REST | | | |
|---|---|---|---|---|
| | 2.0 h void | | 4.8 h void | |
| Organ | rads/ 30 mCi | mGy/ 1,110 MBq | rads/ 30 mCi | mGy/ 1,110 MBq |
| Breasts | 0.2 | 2.0 | 0.2 | 1.9 |
| Gallbladder wall | 2.0 | 20.0 | 2.0 | 20.0 |
| Small intestine | 3.0 | 30.0 | 3.0 | 30.0 |
| Upper large intestine wall | 5.4 | 55.5 | 5.4 | 55.5 |
| Lower large intestine wall | 3.9 | 40.0 | 4.2 | 41.1 |
| Stomach wall | 0.6 | 6.1 | 0.6 | 5.8 |
| Heart wall | 0.5 | 5.1 | 0.5 | 4.9 |
| Kidneys | 2.0 | 20.0 | 2.0 | 20.0 |
| Liver | 0.6 | 5.8 | 0.6 | 5.7 |
| Lungs | 0.3 | 2.8 | 0.3 | 2.7 |
| Bone surfaces | 0.7 | 6.8 | 0.7 | 6.4 |
| Thyroid | 0.7 | 7.0 | 0.7 | 7.0 |
| Ovaries | 1.5 | 15.5 | 1.6 | 15.5 |
| Testes | 0.3 | 3.4 | 0.4 | 3.9 |
| Red marrow | 0.5 | 5.1 | 0.5 | 5.0 |
| Urinary bladder wall | 2.0 | 20.0 | 4.2 | 41.1 |
| Total body | 0.5 | 4.8 | 0.5 | 4.8 |

| | STRESS | | | |
|---|---|---|---|---|
| | 2.0 h void | | 4.8 h void | |
| Organ | rads/ 30 mCi | mGy/ 1,110 MBq | rads/ 30 mCi | mGy/ 1,110 MBq |
| Breasts | 0.2 | 2.0 | 0.2 | 1.8 |
| Gallbladder wall | 2.8 | 28.9 | 2.8 | 27.8 |
| Small intestine | 2.4 | 24.4 | 2.4 | 24.4 |
| Upper large intestine wall | 4.5 | 44.4 | 4.5 | 44.4 |
| Lower large intestine wall | 3.3 | 32.2 | 3.3 | 32.2 |
| Stomach wall | 0.6 | 5.3 | 0.5 | 5.2 |
| Heart wall | 0.5 | 5.6 | 0 5 | 5.3 |
| Kidneys | 1.7 | 16.7 | 1.7 | 16.7 |
| Liver | 0.4 | 4.2 | 0.4 | 4.1 |
| Lungs | 0.3 | 2.6 | 0.2 | 2.4 |
| Bone surfaces | 0.6 | 6.2 | 0.6 | 6.0 |
| Thyroid | 0.3 | 2.7 | 0.2 | 2.4 |
| Ovaries | 1.2 | 12.2 | 1.3 | 13.3 |
| Testes | 0.3 | 3.1 | 0.3 | 3.4 |
| Red marrow | 0.5 | 4.6 | 0.5 | 4.4 |
| Urinary bladder wall | 1.5 | 15.5 | 3.0 | 30.0 |
| Total body | 0.4 | 4.2 | 0.4 | 4.2 |

Radiation dosimetry calculations performed by Radiation Internal Dose Information Center, Oak Ridge Institute for Science and Education, PO Box 117, Oak Ridge, TN 37831-0117.

**Figure 5.15 (Continued)** Sections of a sample radiopharmaceutical package insert. (From Cardiolite, http://www.nuclearonline.org/PI/Cardiolite.pdf)

radiopharmaceutical is approved, little attention is paid to the dosimetry. There is no need to think about individualized dosimetry for each patient, except in the case of a misadministration. In the United States, for example, the Nuclear Regulatory Commission states that this is reportable if:

■ A dose differs from the prescribed dose, or dose that would have resulted from the prescribed dosage, by more than 0.05 Sv (5 rem) effective dose equivalent, 0.5 Sv (50 rem) to an organ or tissue, or 0.5 Sv (50 rem) shallow dose equivalent to the skin, and
  – The total dose delivered differs from the prescribed dose by 20% or more.
  – The total dosage delivered differs from the prescribed dosage by 20% or more or falls outside the prescribed dosage range.
  – The fractionated dose delivered differs from the prescribed dose, for a single fraction, by 50% or more.
■ A dose that exceeds 0.05 Sv (5 rem) effective dose equivalent, 0.5 Sv (50 rem) to an organ or tissue, or 0.5 Sv (50 rem) shallow dose equivalent to the skin from any of the following:
  – An administration of a wrong radioactive drug containing byproduct material
  – An administration of a radioactive drug containing byproduct material by the wrong route of administration
  – An administration of a dose or dosage to the wrong individual or human research subject
  – An administration of a dose or dosage delivered by the wrong mode of treatment
  – A leaking sealed source.
■ A dose to the skin or an organ or tissue other than the treatment site that exceeds by 0.5 Sv (50 rem) and 50% or more of the dose expected from the administration

defined in the written directive (excluding, for permanent implants, seeds that were implanted in the correct site but migrated outside the treatment site).

Also, a report needs to be made to the NRC if any dose to the embryo/fetus or nursing child exceeds 50 mSv.

The International Atomic Energy Agency suggests that a misadministration has occurred if:

- Activity was administered to the wrong patient.
- The wrong radiopharmaceutical or wrong amount of activity (25% more or less than prescribed) was given.
- An unjustified examination was performed on a pregnant or lactating patient.
- An incorrect route of administration was used, including complete extravasation of the injectate.

Another circumstance in which dose calculations will be performed for a diagnostic radiopharmaceutical is in the case of administration to a pregnant, potentially pregnant, or lactating woman. Often, such administrations will be accidental. Most centers rely on a combination of signage (e.g., Figure 5.16) and communication between staff and

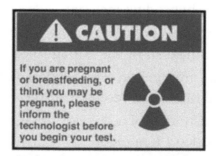

**Figure 5.16 Example sign for a nuclear medicine department. (From Biodex, http://1.bp.blogspot.com/-kUCxDELN8kk/TlkSh6JUkYI/ AAAAAAAAENg/OzZyn9R8dRo/s1600/Radiation-Precautions-In-Pregnancy-Sign-S-8162.gif)**

patients to avoid inadvertent administrations in these cases. Female patients may not be aware that they are pregnant, or may deny it for various reasons. However, intentional administrations occur as well. A complication of pregnancy can be blood clotting, and a nuclear medicine ventilation/perfusion lung scan is often used to diagnose this possibility (Keenan 1992). An extensive study of the placental crossover and fetal radiation dose to the fetus was performed by Russell, Stabin, and Sparks (1997a) and Russell et al. (1997b). Their results are shown in Table 5.2.

Of particular concern is the situation in which radioiodine is administered to a woman who has passed about 10–13 weeks of gestation; the fetal thyroid will have been formed, and this tiny organ concentrates the iodine that crosses the placenta. Evelyn Watson calculated doses to the fetal thyroid per unit activity administered to the mother (Watson 1992). Her results are presented in Table 5.3.

If the pregnant patient is hyperthyroid, her iodine kinetics will differ from the normal case. Stabin et al. (1991) estimated fetal doses for several levels of assumed hyperthyroidism, assuming either very rapid or normal rates of thyroid uptake.

If a pregnant subject had previously had a thyroidectomy, there may be a remnant of thyroid tissue, and/or some thyroid cancer metastases around the body, but usually a large amount of activity is given (enough to destroy all remaining thyroid tissue and the metastases). In a study involving a few athyroidic subjects (Rodriguez 1996), it was found that the kinetics could be well characterized by treating the iodine not taken up by the thyroid by the normal kinetics of urinary bladder excretion (6.1 h half-time). Using these assumptions, and assuming that the other normal soft tissue uptakes occur, and using Russell et al. (1997b) results for fetal uptake and kinetics, they obtained the dose estimates given in Tables 5.4 and 5.5.

**Table 5.2  Fetal Dose Estimates of Russell et al. (1997b)**

| Radiopharmaceutical | Early mGy/MBq | 3-Month mGy/MBq | 6-Month mGy/MBq | 9-Month mGy/MBq |
|---|---|---|---|---|
| $^{57}$Co vitamin B-1, normal—flushing | $1.0 \times 10^{0}$ | $6.8 \times 10^{-1}$ | $8.4 \times 10^{-1}$ | $8.8 \times 10^{-1}$ |
| $^{57}$Co vitamin B-12, normal—no flushing | $1.5 \times 10^{0}$ | $1.0 \times 10^{0}$ | $1.2 \times 10^{0}$ | $1.3 \times 10^{0}$ |
| $^{57}$Co vitamin B-12, PA—flushing | $2.1 \times 10^{-1}$ | $1.7 \times 10^{-1}$ | $1.7 \times 10^{-1}$ | $1.5 \times 10^{-1}$ |
| $^{57}$Co vitamin B-12, PA—no flushing | $2.8 \times 10^{-1}$ | $2.1 \times 10^{-1}$ | $2.2 \times 10^{-1}$ | $2.0 \times 10^{-1}$ |
| $^{58}$Co vitamin B-12, normal—flushing | $2.5 \times 10^{0}$ | $1.9 \times 10^{0}$ | $2.1 \times 10^{0}$ | $2.1 \times 10^{0}$ |
| $^{58}$Co vitamin B-12, normal—no flushing | $3.7 \times 10^{0}$ | $2.8 \times 10^{0}$ | $3.1 \times 10^{0}$ | $3.1 \times 10^{0}$ |
| $^{58}$Co vitamin B-12, PA—flushing | $8.3 \times 10^{-1}$ | $7.4 \times 10^{-1}$ | $6.4 \times 10^{-1}$ | $4.8 \times 10^{-1}$ |
| $^{58}$Co vitamin B-12, PA—no flushing | $9.8 \times 10^{-1}$ | $8.5 \times 10^{-1}$ | $7.6 \times 10^{-1}$ | $6.0 \times 10^{-1}$ |
| $^{60}$Co vitamin B-12, normal—flushing | $3.7 \times 10^{1}$ | $2.8 \times 10^{1}$ | $3.1 \times 10^{1}$ | $3.2 \times 10^{1}$ |
| $^{60}$Co vitamin B-12, normal—no flushing | $5.5 \times 10^{1}$ | $4.2 \times 10^{1}$ | $4.7 \times 10^{1}$ | $4.7 \times 10^{1}$ |
| $^{60}$Co vitamin B-12, PA—flushing | $5.9 \times 10^{0}$ | $4.7 \times 10^{0}$ | $4.8 \times 10^{0}$ | $4.5 \times 10^{0}$ |
| $^{60}$Co vitamin B-12, PA—no flushing | $8.3 \times 10^{0}$ | $6.5 \times 10^{0}$ | $6.8 \times 10^{0}$ | $6.5 \times 10^{0}$ |
| $^{18}$F FDG | $\mathbf{2.2 \times 10^{-2}}$ | $\mathbf{2.2 \times 10^{-2}}$ | $\mathbf{1.7 \times 10^{-2}}$ | $\mathbf{1.7 \times 10^{-2}}$ |
| $^{18}$F sodium fluoride | $2.2 \times 10^{-2}$ | $1.7 \times 10^{-2}$ | $7.5 \times 10^{-3}$ | $6.8 \times 10^{-3}$ |
| $^{67}$Ga citrate | $\mathbf{9.3 \times 10^{-2}}$ | $\mathbf{2.0 \times 10^{-1}}$ | $\mathbf{1.8 \times 10^{-1}}$ | $\mathbf{1.3 \times 10^{-1}}$ |
| $^{123}$I hippuran | $3.1 \times 10^{-2}$ | $2.4 \times 10^{-2}$ | $8.4 \times 10^{-3}$ | $7.9 \times 10^{-3}$ |
| $^{123}$I IMP | $1.9 \times 10^{-2}$ | $1.1 \times 10^{-2}$ | $7.1 \times 10^{-3}$ | $5.9 \times 10^{-3}$ |

*(Continued)*

**Table 5.2 *(Continued)*  Fetal Dose Estimates of Russell et al. (1997b)**

| Radiopharmaceutical | *Early mGy/ MBq* | *3-Month mGv/ MBq* | *6-Month mGy/ MBq* | *9-Month mGy/ MBq* |
|---|---|---|---|---|
| [123]I MIBG | $1.8 \times 10^{-2}$ | $1.2 \times 10^{-2}$ | $6.8 \times 10^{-3}$ | $6.2 \times 10^{-3}$ |
| [124]I sodium iodide | $\mathbf{2.0 \times 10^{-2}}$ | $\mathbf{1.4 \times 10^{-2}}$ | $\mathbf{1.1 \times 10^{-2}}$ | $\mathbf{9.8 \times 10^{-3}}$ |
| [124]I sodium iodide | $\mathbf{1.4 \times 10^{-1}}$ | $\mathbf{1.0 \times 10^{-1}}$ | $\mathbf{5.9 \times 10^{-2}}$ | $\mathbf{4.6 \times 10^{-2}}$ |
| [125]I HSA | $2.5 \times 10^{-1}$ | $7.8 \times 10^{-2}$ | $3.8 \times 10^{-2}$ | $2.6 \times 10^{-2}$ |
| [125]I IMP | $3.2 \times 10^{-2}$ | $1.3 \times 10^{-2}$ | $4.8 \times 10^{-3}$ | $3.6 \times 10^{-3}$ |
| [125]I MIBG | $2.6 \times 10^{-2}$ | $1.1 \times 10^{-2}$ | $4.1 \times 10^{-3}$ | $3.4 \times 10^{-3}$ |
| [125]I sodium iodide | $\mathbf{1.8 \times 10^{-2}}$ | $\mathbf{9.5 \times 10^{-3}}$ | $\mathbf{3.5 \times 10^{-3}}$ | $\mathbf{2.3 \times 10^{-3}}$ |
| [126]I sodium iodide | $\mathbf{7.8 \times 10^{-2}}$ | $\mathbf{5.1 \times 10^{-2}}$ | $\mathbf{3.2 \times 10^{-2}}$ | $\mathbf{2.6 \times 10^{-2}}$ |
| [130]I sodium iodide | $\mathbf{1.8 \times 10^{-1}}$ | $\mathbf{1.3 \times 10^{-1}}$ | $\mathbf{7.6 \times 10^{-2}}$ | $\mathbf{5.7 \times 10^{-2}}$ |
| [131]I hippuran | $6.4 \times 10^{-2}$ | $5.0 \times 10^{-2}$ | $1.9 \times 10^{-2}$ | $1.8 \times 10^{-2}$ |
| [131]I HSA | $5.2 \times 10^{-1}$ | $1.8 \times 10^{-1}$ | $1.6 \times 10^{-1}$ | $1.3 \times 10^{-1}$ |
| [131]I MAA | $6.7 \times 10^{-2}$ | $4.2 \times 10^{-2}$ | $4.0 \times 10^{-2}$ | $4.2 \times 10^{-2}$ |
| [131]I MIBG | $1.1 \times 10^{-1}$ | $5.4 \times 10^{-2}$ | $3.8 \times 10^{-2}$ | $3.5 \times 10^{-2}$ |
| [131]I sodium iodide | $\mathbf{7.2 \times 10^{-2}}$ | $\mathbf{6.8 \times 10^{-2}}$ | $\mathbf{2.3 \times 10^{-1}}$ | $\mathbf{2.7 \times 10^{-1}}$ |
| [131]I rose bengal | $2.2 \times 10^{-1}$ | $2.2 \times 10^{-1}$ | $1.6 \times 10^{-1}$ | $9.0 \times 10^{-2}$ |
| [111]In DTPA | $6.5 \times 10^{-2}$ | $4.8 \times 10^{-2}$ | $2.0 \times 10^{-2}$ | $1.8 \times 10^{-2}$ |
| [111]In pentetreotide | $8.2 \times 10^{-2}$ | $6.0 \times 10^{-2}$ | $3.5 \times 10^{-2}$ | $3.1 \times 10^{-2}$ |
| [111]In platelets | $1.7 \times 10^{-1}$ | $1.1 \times 10^{-1}$ | $9.9 \times 10^{-2}$ | $8.9 \times 10^{-2}$ |
| [111]In red blood cells | $2.2 \times 10^{-1}$ | $1.3 \times 10^{-1}$ | $1.1 \times 10^{-1}$ | $8.6 \times 10^{-2}$ |
| [111]In white blood cells | $1.3 \times 10^{-1}$ | $9.6 \times 10^{-2}$ | $9.6 \times 10^{-2}$ | $9.4 \times 10^{-2}$ |
| [99m]Tc albumin microspheres | $4.1 \times 10^{-3}$ | $3.0 \times 10^{-3}$ | $2.5 \times 10^{-3}$ | $2.1 \times 10^{-3}$ |
| [99m]Tc disofenin | $1.7 \times 10^{-2}$ | $1.5 \times 10^{-2}$ | $1.2 \times 10^{-2}$ | $6.7 \times 10^{-3}$ |
| [99m]Tc DMSA | $\mathbf{5.1 \times 10^{-3}}$ | $\mathbf{4.7 \times 10^{-3}}$ | $\mathbf{4.0 \times 10^{-3}}$ | $\mathbf{3.4 \times 10^{-3}}$ |
| [99m]Tc DTPA | $\mathbf{1.2 \times 10^{-2}}$ | $\mathbf{8.7 \times 10^{-3}}$ | $\mathbf{4.1 \times 10^{-3}}$ | $\mathbf{4.7 \times 10^{-3}}$ |
| [99m]Tc DTPA aerosol | $\mathbf{5.8 \times 10^{-3}}$ | $\mathbf{4.3 \times 10^{-3}}$ | $\mathbf{2.3 \times 10^{-3}}$ | $\mathbf{3.0 \times 10^{-3}}$ |

*(Continued)*

**Table 5.2** *(Continued)*  **Fetal Dose Estimates of Russell et al. (1997b)**

| Radiopharmaceutical | Early mGy/ MBq | 3-Month mGv/ MBq | 6-Month mGy/ MBq | 9-Month mGy/ MBq |
|---|---|---|---|---|
| $^{99m}$Tc glucoheptonate | $1.2 \times 10^{-2}$ | $1.1 \times 10^{-2}$ | $5.3 \times 10^{-3}$ | $4.6 \times 10^{-3}$ |
| $^{99m}$Tc HDP | $5.2 \times 10^{-3}$ | $5.4 \times 10^{-3}$ | $3.0 \times 10^{-3}$ | $2.5 \times 10^{-3}$ |
| $^{99m}$Tc HEDP | $7.2 \times 10^{-3}$ | $5.2 \times 10^{-3}$ | $2.7 \times 10^{-3}$ | $2.4 \times 10^{-3}$ |
| $^{99m}$Tc HMPAO | $8.7 \times 10^{-3}$ | $6.7 \times 10^{-3}$ | $4.8 \times 10^{-3}$ | $3.6 \times 10^{-3}$ |
| $^{99m}$Tc human serum albumin | $5.1 \times 10^{-3}$ | $3.0 \times 10^{-3}$ | $2.6 \times 10^{-3}$ | $2.2 \times 10^{-3}$ |
| $^{99m}$Tc MAA | $2.8 \times 10^{-3}$ | $4.0 \times 10^{-3}$ | $5.0 \times 10^{-3}$ | $4.0 \times 10^{-3}$ |
| $^{99m}$Tc MAG3 | $1.8 \times 10^{-2}$ | $1.4 \times 10^{-2}$ | $5.5 \times 10^{-3}$ | $5.2 \times 10^{-3}$ |
| $^{99m}$Tc MDP | $6.1 \times 10^{-3}$ | $5.4 \times 10^{-3}$ | $2.7 \times 10^{-3}$ | $2.4 \times 10^{-3}$ |
| $^{99m}$Tc MIBI—rest | $1.5 \times 10^{-2}$ | $1.2 \times 10^{-2}$ | $8.4 \times 10^{-3}$ | $5.4 \times 10^{-3}$ |
| $^{99m}$Tc MIBI—stress | $1.2 \times 10^{-2}$ | $9.5 \times 10^{-3}$ | $6.9 \times 10^{-3}$ | $4.4 \times 10^{-3}$ |
| $^{99m}$Tc pertechnetate | $1.1 \times 10^{-2}$ | $2.2 \times 10^{-2}$ | $1.4 \times 10^{-2}$ | $9.3 \times 10^{-3}$ |
| $^{99m}$Tc PYP | $6.0 \times 10^{-3}$ | $6.6 \times 10^{-3}$ | $3.6 \times 10^{-3}$ | $2.9 \times 10^{-3}$ |
| $^{99m}$Tc RBC—heat treated | $1.7 \times 10^{-3}$ | $1.6 \times 10^{-3}$ | $2.1 \times 10^{-3}$ | $2.2 \times 10^{-3}$ |
| $^{99m}$Tc RBC—*in vitro* | $6.8 \times 10^{-3}$ | $4.7 \times 10^{-3}$ | $3.4 \times 10^{-3}$ | $2.8 \times 10^{-3}$ |
| $^{99m}$Tc RBC—*in vivo* | $6.4 \times 10^{-3}$ | $4.3 \times 10^{-3}$ | $3.3 \times 10^{-3}$ | $2.7 \times 10^{-3}$ |
| $^{99m}$Tc sulfur colloid—normal | $1.8 \times 10^{-3}$ | $2.1 \times 10^{-3}$ | $3.2 \times 10^{-3}$ | $3.7 \times 10^{-3}$ |
| $^{99m}$Tc sulfur colloid—liver disease | $3.2 \times 10^{-3}$ | $2.5 \times 10^{-3}$ | $2.8 \times 10^{-3}$ | $2.8 \times 10^{-3}$ |
| $^{99m}$Tc teboroxime | $8.9 \times 10^{-3}$ | $7.1 \times 10^{-3}$ | $5.8 \times 10^{-3}$ | $3.7 \times 10^{-3}$ |
| $^{99m}$Tc tetrofosmin | $9.6 \times 10^{-3}$ | $7.0 \times 10^{-3}$ | $5.4 \times 10^{-3}$ | $3.6 \times 10^{-3}$ |
| $^{99m}$Tc white blood cells | $3.8 \times 10^{-3}$ | $2.8 \times 10^{-3}$ | $2.9 \times 10^{-3}$ | $2.8 \times 10^{-3}$ |
| $^{201}$Tl chloride | $9.7 \times 10^{-2}$ | $5.8 \times 10^{-2}$ | $4.7 \times 10^{-2}$ | $2.7 \times 10^{-2}$ |
| $^{127}$Xe, 5-minute rebreathing, 5-liter spirometer volume | $4.3 \times 10^{-4}$ | $2.4 \times 10^{-4}$ | $1.9 \times 10^{-4}$ | $1.5 \times 10^{-4}$ |

*(Continued)*

**Table 5.2** *(Continued)* **Fetal Dose Estimates of Russell et al. (1997b)**

| Radiopharmaceutical | Early mGy/ MBq | 3-Month mGv/ MBq | 6-Month mGy/ MBq | 9-Month mGy/ MBq |
|---|---|---|---|---|
| $^{127}$Xe, 5-minute rebreathing, 7.5-liter spirometer volume | $2.3 \times 10^{-4}$ | $1.3 \times 10^{-4}$ | $1.0 \times 10^{-4}$ | $8.4 \times 10^{-5}$ |
| $^{127}$Xe, 5-minute rebreathing, 10-liter spirometer volume | $2.3 \times 10^{-4}$ | $1.4 \times 10^{-4}$ | $1.1 \times 10^{-4}$ | $9.2 \times 10^{-5}$ |
| $^{133}$Xe, 5-minute rebreathing, 5-liter spirometer volume | $4.1 \times 10^{-4}$ | $4.8 \times 10^{-5}$ | $3.5 \times 10^{-5}$ | $2.6 \times 10^{-5}$ |
| $^{133}$Xe, 5-minute rebreathing, 7.5-liter spirometer volume | $2.2 \times 10^{-4}$ | $2.6 \times 10^{-5}$ | $1.9 \times 10^{-5}$ | $1.5 \times 10^{-5}$ |
| $^{133}$Xe, 5-minute rebreathing. 10-liter spirometer volume | $2.5 \times 10^{-4}$ | $2.9 \times 10^{-5}$ | $2.1 \times 10^{-5}$ | $1.6 \times 10^{-5}$ |
| $^{133}$Xe, injection | $4.9 \times 10^{-6}$ | $1.0 \times 10^{-6}$ | $1.4 \times 10^{-6}$ | $1.6 \times 10^{-6}$ |

**Table 5.3** **Fetal Thyroid Doses (mGy to the Fetal Thyroid per MBq Administered to the Mother)**

| Gestational Age (months) | I-123 | I-124 | I-125 | I-131 |
|---|---|---|---|---|
| 3 | 2.7 | 24 | 290 | 230 |
| 4 | 2.6 | 27 | 240 | 260 |
| 5 | 6.4 | 76 | 280 | 580 |
| 6 | 6.4 | 100 | 210 | 550 |
| 7 | 4.1 | 96 | 160 | 390 |
| 8 | 4.0 | 110 | 150 | 350 |
| 9 | 2.9 | 99 | 120 | 270 |

*Source:* From Watson EE, Radiation absorbed dose to the human fetal thyroid, In *Fifth International Radiopharmaceutical Dosimetry Symposium*, Oak Ridge Associated Universities, Oak Ridge, TN, 1992, pp. 179–187.

**Table 5.4   Fetal Doses (Early Pregnancy) for a Hyperthyroid Pregnant Subject (mGy to the Fetus per MBq Administered to the Mother)**

| *Maximum* thyroid uptake | 20% | 40% | 60% | 80% | 100% |
|---|---|---|---|---|---|
| *Fast* thyroid uptake | 0.049 | 0.044 | 0.040 | 0.036 | 0.036 |
| *Normal* thyroid uptake | 0.063 | 0.058 | 0.055 | 0.052 | 0.053 |

*Source:* From Stabin MG et al., 1991, *J Nucl Med*, 32, 808–813.

**Table 5.5   Fetal Thyroid Doses for Athyroid Pregnant Subjects (mGy to the Fetus per MBq Administered to the Mother)**

| *Stage of Pregnancy* | *Fetal Dose* |
|---|---|
| Early pregnancy | 0.068 |
| 3 months | 0.070 |
| 6 months | 0.225 |
| 9 months | 0.27 |

*Source:* From Rodriguez M, 1996, *Development of a Kinetic Model and Calculation of Radiation Dose Estimates Forsodium-Iodide-[131]I in Athyroid Individuals.* Masters Project, Colorado State University.

An unusual case is when radioiodine is administered to a subject who *subsequently* becomes pregnant. Some of the iodine will have washed out of her system, the quantity depending on the time between her study and when she became pregnant. Sparks and Stabin (1996) treated these unusual kinetics and developed the dose estimates given in Tables 5.6 (hyperthyroid cases) and 5.7 (euthyroid cases).

**Table 5.6   Dose to the Fetus when Conception Occurs after Administration of I-131, Hyperthyroid Cases (mGy to the Fetus per MBq Administered to the Mother)**

| % Max Uptake | Time in Weeks after Administration that Conception Occurs | | | | | | | |
|---|---|---|---|---|---|---|---|---|
| | *1* | *2* | *3* | *4* | *5* | *6* | *7* | *8* |
| 5 | 4.1E–04 | 1.9E–04 | 8.7E–05 | 4.0E–05 | 1.9E–05 | 8.7E–06 | 4.0E–06 | 1.9E–06 |
| 10 | 8.3E–04 | 3.8E–04 | 1.7E–04 | 8.0E–05 | 3.7E–05 | 1.7E–05 | 7.8E–06 | 3.6E–06 |
| 15 | 1.3E–03 | 5.8E–04 | 2.6E–04 | 1.2E–04 | 5.5E–05 | 2.5E–05 | 1.1E–05 | 5.2E–06 |
| 20 | 1.7E–03 | 7.8E–04 | 3.5E–04 | 1.6E–04 | 7.2E–05 | 3.3E–05 | 1.5E–05 | 6.7E–06 |
| 25 | 2.2E–03 | 9.8E–04 | 4.4E–04 | 2.0E–04 | 8.8E–05 | 4.0E–05 | 1.8E–05 | 8.0E–06 |
| 30 | 2.7E–03 | 1.2E–03 | 5.3E–04 | 2.3E–04 | 1.0E–04 | 4.6E–05 | 2.0E–05 | 9.1E–06 |
| 35 | 3.2E–03 | 1.4E–03 | 6.1E–04 | 2.7E–04 | 1.2E–04 | 5.2E–05 | 2.3E–05 | 1.0E–05 |
| 40 | 3.7E–03 | 1.6E–03 | 7.0E–04 | 3.0E–04 | 1.3E–04 | 5.7E–05 | 2.4E–05 | 1.1E–05 |
| 45 | 4.3E–03 | 1.8E–03 | 7.8E–04 | 3.3E–04 | 1.4E–04 | 6.0E–05 | 2.6E–05 | 1.1E–05 |
| 50 | 4.8E–03 | 2.0E–03 | 8.5E–04 | 3.6E–04 | 1.5E–04 | 6.3E–05 | 2.6E–05 | 1.1E–05 |
| 55 | 5.4E–03 | 2.2E–03 | 9.2E–04 | 3.8E–04 | 1.6E–04 | 6.4E–05 | 2.7E–05 | 1.1E–05 |
| 60 | 6.0E–03 | 2.4E–03 | 9.8E–04 | 4.0E–04 | 1.6E–04 | 6.4E–05 | 2.6E–05 | 1.0E–05 |
| 65 | 6.7E–03 | 2.6E–03 | 1.0E–03 | 4.0E–04 | 1.6E–04 | 6.2E–05 | 2.5E–05 | 9.7E–06 |
| 70 | 7.3E–03 | 2.8E–03 | 1.1E–03 | 4.1E–04 | 1.5E–04 | 5.9E–05 | 2.2E–05 | 8.6E–06 |
| 75 | 7.9E–03 | 2.9E–03 | 1.1E–03 | 4.0E–04 | 1.5E–04 | 5.4E–05 | 2.0E–05 | 7.2E–06 |
| 80 | 8.5E–03 | 3.0E–03 | 1.1E–03 | 3.7E–04 | 1.3E–04 | 4.6E–05 | 1.6E–05 | 5.7E–06 |
| 85 | 9.1E–03 | 3.0E–03 | 1.0E–03 | 3.4E–04 | 1.1E–04 | 3.8E–05 | 1.3E–05 | 4.2E–06 |
| 90 | 9.6E–03 | 3.0E–03 | 9.2E–04 | 2.9E–04 | 8.9E–05 | 2.8E–05 | 8.6E–06 | 2.7E–06 |
| 95 | 9.8E–03 | 2.8E–03 | 7.9E–04 | 2.2E–04 | 6.3E–05 | 1.8E–05 | 5.1E–06 | 1.4E–06 |
| 100 | 9.8E–03 | 2.4E–03 | 6.1E–04 | 1.5E–04 | 3.8E–05 | 9.3E–06 | 2.3E–06 | 5.8E–07 |

**Table 5.7  Dose to the Fetus when Conception Occurs after Administration of I-131, Euthyroid Cases (mGy to the Fetus per MBq Administered to the Mother)**

| % Max Uptake | Time in Weeks after Administration that Conception Occurs | | | | | | | |
|---|---|---|---|---|---|---|---|---|
| | 1 | 2 | 3 | 4 | 5 | 6 | 7 | 8 |
| 5 | 3.1E–04 | 1.5E–04 | 7.7E–05 | 3.8E–05 | 1.9E–05 | 9.5E–06 | 4.7E–06 | 2.4E–06 |
| 15 | 8.8E–04 | 4.4E–04 | 2.2E–04 | 1.1E–04 | 5.6E–05 | 2.8E–05 | 1.4E–05 | 7.2E–06 |
| 25 | 1.4E–03 | 7.1E–04 | 3.6E–04 | 1.8E–04 | 9.2E–05 | 4.7E–05 | 2.4E–05 | 1.2E–05 |

# Summary

Performing dose calculations for diagnostic radiopharmaceuticals depends on gathering and analyzing a significant amount of data, whether relying on extrapolated animal data or human data, both of which are required for regulatory approval of a new agent. Once the appropriate number of data points is obtained and a suitable kinetic analysis is performed, one has the numbers of disintegrations in all important source organs and needs to only enter this in the OLINDA/EXM code (Stabin, Sparks, and Crowe 2005) to obtain dose estimates in all target organs and the effective dose. Except for the few cases noted in the previous section, these tables of reference doses are summarized in dose estimate compendia and rarely referenced. During a shortage of $^{99m}$Tc between 2008 and 2010, replacement radiopharmaceuticals were sought to perform the studies routinely done with this very popular radionuclide. In those times, effective doses for some candidate radiopharmaceuticals were compared to help choose candidates, so dose estimate tables became relevant. Dose estimate tables are frequently updated, as metabolic models for radiopharmaceuticals are re-evaluated, so printed and online compendia need to be maintained as up-to-date as possible; this is the task of a number of groups, including the RAdiation Dose Assessment Resource Task Group of the Society of Nuclear Medicine and Molecular Imaging.

# Bibliography

Bodei L, Cremonesi M, Grana CM, Fazio N, Iodice S, Baio SM, Bartolomei M, et al. 2011. Peptide receptor radionuclide therapy with [177]Lu-DOTATATE: the IEO phase I–II study. *Eur J Nucl Med Mol Imaging*, 38, 2125–2135.

Breitz HB, Wendt III RE, Stabin MG, Shen S, Erwin WD, Rajendran JG, Eary JF, et al. 2006. [166]Ho-DOTMP radiation-absorbed dose estimation for skeletal targeted radiotherapy. *J Nucl Med*, 47, 534–542.

Dewaraja YK, Frey EC, Sgouros G, Brill AB, Roberson P, Zanzonico PB, and Ljungberg M. 2012. MIRD Pamphlet No. 23: quantitative SPECT for patient-specific 3-dimensional dosimetry in internal radionuclide therapy. *J Nucl Med*, 53(8), 1310–1325.

Drozdovitch V, Brill AB, Callahan RJ, Clanton JA, DePietro A, Goldsmith SJ, Greenspan BS, et al. 2015. Use of radiopharmaceuticals in diagnostic nuclear medicine in the United States: 1960–2010. *Health Phys*, 108(5), 520–537.

Gambhir S, Mahoney DK, Turner MS, Wong ATC, Rosenqvist G, Huang SC, and Phelps ME. 1995. Symbolic interactive modeling package and learning environment (SIMPLE): a new compartmental modeling tool. In *Simulation in the Medical Sciences*, edited by JG Anderson and M Katzper (pp. 173–186). The Society for Computer Simulation, Phoenix 5, San Diego, CA.

Keenan AM. 1992. Nuclear medicine in the diagnosis of deep venous thrombosis and pulmonary embolism. *Semin Arthroplasty*, 3(2), 84–94.

Pereira JM, Stabin MG, Lima FRA, Guimaraes MICC, and Forrester JW. 2010. Image quantification for radiation dose calculations—limitations and uncertainties. *Health Phys*, 99(5), 688–701.

Rodriguez M. 1996. *Development of a Kinetic Model and Calculation of Radiation Dose Estimates For sodium-Iodide-[131]I in Athyroid Individuals*. Masters Project, Colorado State University.

Russell JR, Stabin MG, and Sparks RB. 1997a. Placental transfer of radiopharmaceuticals and dosimetry in pregnancy. *Health Phys*, 73(5), 747–755.

Russell JR, Stabin MG, Sparks RB, and Watson EE. 1997b. Radiation absorbed dose to the embryo/fetus from radiopharmaceuticals. *Health Phys*, 73(5), 756–769.

Siegel J, Thomas S, Stubbs J, Stabin M, Hays M, Koral K, Robertson J, et al. 1999. MIRD Pamphlet No 16: techniques for quantitative radiopharmaceutical biodistribution data acquisition and analysis for use in human radiation dose estimates. *J Nucl Med*, 40, 37S–61S.

Siegel JA, Stabin MG, and Brill AB. 2002. The importance of patient-specific radiation dose calculations for the administration of radionuclides in therapy. *Cell Mol Biol* (Noisy-le-grand), 48(5), 451–459.

Sparks RB and Aydogan B. 1999. Comparison of the effectiveness of some common animal data scaling techniques in estimating human radiation dose. In *Sixth International Radiopharmaceutical Dosimetry Symposium*, edited by A Stelson, M Stabin, and R Sparks (pp. 705–716). Oak Ridge Institute for Science and Education, Oak Ridge, TN.

Sparks RB and Stabin MG. 1996. Fetal radiation dose estimates for I-131 sodium iodide in cases where accidental conception occurs after administration. *Presented at the Sixth International Radiopharmaceutical Dosimetry Symposium held May 7–10, 1996 in Gatlinburg, TN*, edited by A Stelson, M Stabin, and R Sparks (pp. 360–364). Oak Ridge Associated Universities, Oaks Ridge, TN.

Stabin MG. 2008. Fundamentals of Nuclear Medicine Dosimetry. Springer, New York, NY.

Stabin MG, Sparks RB, and Crowe E. 2005. OLINDA/EXM: the second-generation personal computer software for internal dose assessment in nuclear medicine. *J Nucl Med*, 46, 1023–1027.

Stabin MG, Watson EE, Marcus CS, and Salk RD. 1991. Radiation dosimetry for the adult female and fetus from iodine-131 administration in hyperthyroidism. *J Nucl Med*, 32, 808–813.

VanBrocklin HF. 2008. Radiopharmaceuticals for Drug Development: United States Regulatory Perspective. *Curr Radiopharm*, 1, 2–6.

Van de Wiele C, Dumont F, Dierckx RA, Peers SH, Thornback JR, Slegers G, and Thierens H. 2001. Biodistribution and dosimetry of $^{99m}$Tc-RP527, a Gastrin-Releasing Peptide (GRP) agonist for the visualization of GRP receptor-expressing malignancies. *J Nucl Med*, 42, 1722–1727.

Vishal P, Rahulgiri G, Pratik M, and Kumar B. 2014. A review on drug approval process for US, Europe, and India. *Int J Drug Regulatory Affairs*, 2(1), 1–11.

Watson EE. 1992. Radiation absorbed dose to the human fetal thyroid. In *Fifth International Radiopharmaceutical Dosimetry Symposium* (pp. 179–187). Oak Ridge Associated Universities, Oak Ridge, TN.

Windhover's *In vivo*: The Business & Medicine Report, Bain drug economics model, Windhover Information Inc, Norwalk, CT, 2003. Available at http://www.bain.com/bainweb/PDFs/cms/Marketing/rebuilding_big_pharma.pdf

Wiseman GA, Kornmehl E, Leigh B, Erwin WD, Podoloff DA, Spies S, Sparks RB, Stabin MG, Witzig T, and White CA. 2003. Radiation dosimetry results and safety correlations from $^{90}$Y-ibritumomab tiuxetan radioimmunotherapy for relapsed or refractory non-Hodgkin's lymphoma: combined data from 4 clinical trials. *J Nucl Med*, 44, 465–474.

# Chapter 6

# Approved Radiopharmaceuticals

# Use of Radiopharmaceuticals (Table 6.1)

**Table 6.1 Use of Radiopharmaceuticals in the United States from 1960 to 2010**

| Radiopharmaceutical | Percentage of Use of Radiopharmaceuticals for Diagnostic Nuclear Medicine Procedures in | | | | | | | | | |
|---|---|---|---|---|---|---|---|---|---|---|
| | 1960–1964 | 1965–1969 | 1970–1974 | 1975–1979 | 1980–1984 | 1985–1989 | 1990–1994 | 1995–1999 | 2000–2004 | 2005–2010 |
| **Thyroid Scan** | | | | | | | | | | |
| $^{131}$I-sodium iodide | 100 | 88 | 53 | 22 | 19 | 14 | 10 | 9 | 8 | 8 |
| $^{123}$I-sodium iodide | –a | – | 3 | 16 | 22 | 26 | 36 | 49 | 51 | 56 |
| $^{99m}$Tc-pertechnetate | – | 12 | 44 | 62 | 59 | 60 | 54 | 42 | 41 | 36 |
| **Thyroid Update** | | | | | | | | | | |
| $^{131}$I-sodium iodide | 100 | 92 | 93 | 76 | 67 | 67 | 60 | 55 | 48 | 45 |
| $^{123}$I-sodium iodide | – | – | – | 13 | 22 | 22 | 38 | 45 | 52 | 55 |
| $^{99m}$Tc-pertechnetate | – | 8 | 7 | 11 | 11 | 11 | 2 | – | – | – |
| **Brain Scan** | | | | | | | | | | |
| $^{201}$Hg-chlormerodrin | 81 | 36 | – | – | – | – | – | – | – | – |

(Continued)

**Table 6.1 (Continued)  Use of Radiopharmaceuticals in the United States from 1960 to 2010**

| Radiopharmaceutical | Percentage of Use of Radiopharmaceuticals for Diagnostic Nuclear Medicine Procedures in | | | | | | | | | |
|---|---|---|---|---|---|---|---|---|---|---|
| | 1960–1964 | 1965–1969 | 1970–1974 | 1975–1979 | 1980–1984 | 1985–1989 | 1990–1994 | 1995–1999 | 2000–2004 | 2005–2010 |
| $^{197}$Hg-chlormerodrin | 19 | 17 | 3 | — | — | — | — | — | — | — |
| $^{99m}$Tc-pertechnetate | — | 47 | 89 | 53 | 33 | 24 | 4 | — | — | — |
| $^{99m}$Tc-DTPA | — | — | 8 | 43 | 56 | 48 | 35 | 25 | 20 | 18 |
| $^{99m}$Tc glucoheptonate | — | — | — | 4 | 11 | 8 | — | — | — | — |
| $^{99m}$Tc-HMPAO | — | — | — | — | — | 13 | 39 | 35 | 20 | 19 |
| $^{99m}$Tc-ECD | — | — | — | — | — | — | 12 | 25 | 37 | 35 |
| $^{201}$Tl-chloride | — | — | — | — | — | 10 | 1 | — | — | — |
| $^{18}$F-FDG | — | — | — | — | — | 4 | 9 | 15 | 23 | 28 |
| **Brain Blood Flow** | | | | | | | | | | |
| $^{131}$I-RIHSA | 100 | 40 | 15 | 3 | — | — | — | — | — | — |
| $^{99m}$Tc-pertechnetate | — | 53 | 63 | 42 | 23 | 19 | 3 | — | — | — |

*(Continued)*

**Table 6.1 (Continued)  Use of Radiopharmaceuticals in the United States from 1960 to 2010**

| Radiopharmaceutical | Percentage of Use of Radiopharmaceuticals for Diagnostic Nuclear Medicine Procedures in | | | | | | | | | |
|---|---|---|---|---|---|---|---|---|---|---|
| | 1960–1964 | 1965–1969 | 1970–1974 | 1975–1979 | 1980–1984 | 1985–1989 | 1990–1994 | 1995–1999 | 2000–2004 | 2005–2010 |
| 99mTc-DTPA | — | 7 | 22 | 55 | 67 | 65 | 50 | 47 | 48 | 48 |
| 99mTc-HMPAO | — | — | — | — | — | 2 | 25 | 19 | 11 | 10 |
| 99mTc-ECD | — | — | — | — | — | — | 2 | 17 | 25 | 26 |
| 111Tc-DTPA | — | — | — | — | 10 | 14 | 20 | 17 | 16 | 16 |
| **Lung Perfusion** | | | | | | | | | | |
| 131I MAA | 100 | 93 | 38 | 4 | — | — | — | — | — | — |
| 99mTc-MAA | — | 7 | 58 | 77 | 81 | 94 | 100 | 100 | 100 | 100 |
| 99mTc-HAM | — | — | 4 | 19 | 19 | 6 | — | — | — | — |
| **Lung Ventilation** | | | | | | | | | | |
| 133Xe | — | 100 | 100 | 100 | 89 | 74 | 67 | 64 | 37 | 32 |
| 99mTc-DTPA aerosol | — | — | — | — | 11 | 26 | 33 | 36 | 63 | 68 |

*(Continued)*

**Table 6.1 (Continued)  Use of Radiopharmaceuticals in the United States from 1960 to 2010**

| Radiopharmaceutical | Percentage of Use of Radiopharmaceuticals for Diagnostic Nuclear Medicine Procedures in | | | | | | | | | |
|---|---|---|---|---|---|---|---|---|---|---|
| | 1960–1964 | 1965–1969 | 1970–1974 | 1975–1979 | 1980–1984 | 1985–1989 | 1990–1994 | 1995–1999 | 2000–2004 | 2005–2010 |
| **Bone Scan** | | | | | | | | | | |
| **Sr-chloride | 100 | 45 | 17 | — | — | — | — | — | — | — |
| **Sr-nitrate | — | 15 | 10 | — | — | — | — | — | — | — |
| Na $^{13}$F | — | 25 | 21 | — | — | — | — | — | — | — |
| $^{99m}$Tc-polyphosphate | — | 15 | 48 | 50 | 4 | — | — | — | — | — |
| $^{99m}$Tc-MDP | — | — | 4 | 50 | 96 | 100 | 100 | 100 | 100 | 100 |
| **Liver Scan** | | | | | | | | | | |
| **Au-colloid | 97 | 47 | 9 | — | — | — | — | — | — | — |
| $^{131}$I-rose bengal | 3 | 18 | — | — | — | — | — | — | — | — |
| $^{99m}$Tc-SC | — | 35 | 91 | 100 | 93 | 82 | 84 | 83 | 79 | 78 |
| $^{99m}$Tc-IDA | — | — | — | — | 7 | 18 | 16 | 17 | 21 | 22 |

*(Continued)*

**Table 6.1 (Continued)** Use of Radiopharmaceuticals in the United States from 1960 to 2010

| Radiopharmaceutical | Percentage of Use of Radiopharmaceuticals for Diagnostic Nuclear Medicine Procedures in | | | | | | | | | |
|---|---|---|---|---|---|---|---|---|---|---|
| | 1960–1964 | 1965–1969 | 1970–1974 | 1975–1979 | 1980–1984 | 1985–1989 | 1990–1994 | 1995–1999 | 2000–2004 | 2005–2010 |
| **Hepatobiliary Scan** | | | | | | | | | | |
| [131]I-rose bengal | – | 100 | 100 | 73 | 16 | – | – | – | – | – |
| [99m]Tc-mebrofenin | – | – | – | 20 | 3 | 4 | 1 | – | – | – |
| [99m]Tc-HIDA | – | – | – | 7 | 76 | 95 | 90 | 84 | 73 | 70 |
| [99m]Tc-DISIDA | – | – | – | – | – | 1 | 9 | 16 | 27 | 30 |
| **Bone Marrow Scan** | | | | | | | | | | |
| [198]Au-colloid | 100 | 65 | 15 | – | – | – | – | – | – | – |
| [99m]Tc-SC | – | 35 | 85 | 100 | 98 | 85 | 93 | 100 | 100 | 100 |
| [99m]Tc-albumin colloid | – | – | – | – | 2 | 15 | 7 | – | – | – |
| Panc**as scan | | | | | | | | | | |
| [75]Se-methionine | 100 | 100 | 80 | 80 | 42 | – | – | – | – | – |

*(Continued)*

**Table 6.1 (Continued)   Use of Radiopharmaceuticals in the United States from 1960 to 2010**

| Radiopharmaceutical | Percentage of Use of Radiopharmaceuticals for Diagnostic Nuclear Medicine Procedures in | | | | | | | | | |
|---|---|---|---|---|---|---|---|---|---|---|
| | 1960–1964 | 1965–1969 | 1970–1974 | 1975–1979 | 1980–1984 | 1985–1989 | 1990–1994 | 1995–1999 | 2000–2004 | 2005–2010 |
| 99mTc-SC | — | — | 20 | 20 | 45 | 30 | 25 | 25 | 25 | 25 |
| 18F-FDG | — | — | — | — | 13 | 70 | 75 | 75 | 75 | 75 |
| **Kidney Scan** | | | | | | | | | | |
| 123I-hippo** | — | 41 | 68 | 34 | 27 | 18 | 2 | — | — | — |
| 203Hg-chlonmerodrin | 56 | 17 | — | — | — | — | — | — | — | — |
| 197Hg-chlonmerodrin | 44 | 42 | 15 | — | — | — | — | — | — | — |
| 99mTc-DTPA | — | — | 16 | 55 | 60 | 61 | 34 | 27 | 5 | 3 |
| 99mTc-gluocheptonate | — | — | 1 | 10 | 7 | 13 | 1 | 5 | 2 | — |
| 99mTc-DMSA | — | — | — | 1 | 6 | 8 | 6 | 9 | 3 | 4 |
| 99mTc-MAG3 | — | — | — | — | — | — | 57 | 59 | 90 | 93 |

*(Continued)*

**Table 6.1 (Continued)   Use of Radiopharmaceuticals in the United States from 1960 to 2010**

| Radiopharmaceutical | Percentage of Use of Radiopharmaceuticals for Diagnostic Nuclear Medicine Procedures in | | | | | | | | | |
|---|---|---|---|---|---|---|---|---|---|---|
| | 1960–1964 | 1965–1969 | 1970–1974 | 1975–1979 | 1980–1984 | 1985–1989 | 1990–1994 | 1995–1999 | 2000–2004 | 2005–2010 |
| **Cardiac Procedures** | | | | | | | | | | |
| $^{201}$Tl-chloride | – | – | 10 | 70 | 67 | 67 | 49 | 30 | 19 | 9 |
| $^{99m}$Tc-pysophosphate | – | 100 | 33 | 12 | 13 | 13 | 7 | – | – | – |
| $^{99m}$Tc-pertechnetate | – | – | 35 | – | – | – | – | – | – | – |
| $^{99m}$Tc-RBC | – | – | 22 | 11 | 20 | 18 | 14 | 13 | 11 | 10 |
| $^{99m}$Tc-HSA | – | – | – | 7 | – | – | – | – | – | – |
| $^{99m}$Tc-sestamibi (1 day) | – | – | – | – | – | 2 | 26 | 47 | 53 | 55 |
| $^{99m}$Tc-sestamibi (2 day) | – | – | – | – | – | – | 2 | 2 | 3 | 4 |
| $^{99m}$Tc-tetrofosmin | – | – | – | – | – | – | 1 | 7 | 13 | 20 |
| $^{13}$F-FDG | – | – | – | – | – | – | 1 | 1 | 1 | 1 |
| $^{112}$Rb-chloride | – | – | – | – | – | – | – | – | – | 1 |

*(Continued)*

**Table 6.1 (Continued)   Use of Radiopharmaceuticals in the United States from 1960 to 2010**

| Radiopharmaceutical | *Percentage of Use of Radiopharmaceuticals for Diagnostic Nuclear Medicine Procedures in* | | | | | | | | | |
|---|---|---|---|---|---|---|---|---|---|---|
| | 1960–1964 | 1965–1969 | 1970–1974 | 1975–1979 | 1980–1984 | 1985–1989 | 1990–1994 | 1995–1999 | 2000–2004 | 2005–2010 |
| **CI Bleeding and Med\*\* Scan** | | | | | | | | | | |
| $^{99m}$Tc-RBC | — | 100 | 96 | 83 | 92 | 89 | 96 | 97 | 97 | 97 |
| $^{99m}$Tc-pertechnetate | — | — | 4 | 17 | 8 | 11 | 4 | 3 | 3 | 3 |
| **Tumor Localization on** | | | | | | | | | | |
| $^{131}$I-HSA | 100 | 100 | 71 | 7 | — | — | — | — | — | — |
| $^{67}$Ga-citrate | — | — | 29 | 93 | 100 | 97 | 84 | 50 | 24 | 1 |
| $^{111}$In-pentetrentide | — | — | — | — | — | — | 1 | 5 | 11 | 11 |
| $^{111}$In-capromab pendetide | — | — | — | — | — | — | — | 3 | 3 | 1 |
| $^{13}$F-FDG | — | — | — | — | — | 3 | 15 | 42 | 62 | 87 |

*(Continued)*

**Table 6.1 (Continued)   Use of Radiopharmaceuticals in the United States from 1960 to 2010**

| Radiopharmaceutical | Percentage of Use of Radiopharmaceuticals for Diagnostic Nuclear Medicine Procedures in | | | | | | | | | |
|---|---|---|---|---|---|---|---|---|---|---|
| | 1960–1964 | 1965–1969 | 1970–1974 | 1975–1979 | 1980–1984 | 1985–1989 | 1990–1994 | 1995–1999 | 2000–2004 | 2005–2010 |
| **Non-Imaging Studies** | | | | | | | | | | |
| *Blood Volume* | | | | | | | | | | |
| 131I-RIHSA | 62 | 60 | 49 | 31 | 42 | 49 | 46 | 60 | 66 | 85 |
| 51Ct-RBC | 38 | 34 | 41 | 57 | 48 | 41 | 46 | 40 | 34 | 17 |
| 125I-HSA | – | 6 | 10 | 12 | 10 | 10 | 8 | – | – | – |
| *Iron Metabolism* | | | | | | | | | | |
| 59Fe-citrate | 100 | 100 | 100 | 100 | 100 | 100 | 100 | – | – | – |

*Source:* From Drozdovitch V, Brill AB, Callahan R J, Clanton JA, DePietro A, Goldsmith SJ, Greenspan BS et al. 2015. Use of Radiopharmaceuticals in Diagnostic Nuclear Medicine in the United States: 1960–2010. *Health Phys,* 108(5), 520–537.

a Use of radiopharmaceutical was not reported for this time period.

# Currently Approved Radiopharmaceuticals (Table 6.2)

**Table 6.2  Currently Approved Radiopharmaceuticals in the United States**

| | | | |
|---|---|---|---|
| 1 | Carbon-11 choline | Mayo Clinic | — | Indicated for positron emission tomography (PET) imaging of patients with suspected prostate cancer recurrence based upon elevated blood prostate-specific antigen (PSA) levels following initial therapy and non-informative bone scintigraphy, computerized tomography (CT) or magnetic resonance imaging (MRI) to help identify potential sites of prostate cancer recurrence for subsequent histologic confirmation |
| 2 | Carbon-14 urea | Kimberly-Clark | PYtest | Detection of gastric urease as an aid in the diagnosis of *H. pylori* infection in the stomach |
| 3 | Fluorine-18 florbetaben | Piramal Imaging | Neuraceq™ | Indicated for PET imaging of the brain to estimate β-amyloid neuritic plaque density in adult patients with cognitive impairment who are being evaluated for Alzheimer's disease (AD) or other causes of cognitive decline |
| 4 | Fluonne-18 florbetapir | Bi Lilly | Amyvid™ | |

(Continued)

**Table 6.2 (Continued)  Currently Approved Radiopharmaceuticals in the United States**

| 5 | Fluorine-18 sodium fluoride[1] | Various | — | PET bone imaging agent to delineate areas of altered osteogenesis |
| 6 | Fluorine-18 fludeoxyglucose[1] | Various | — | As a PET imaging agent to:<br>• Assess abnormal glucose metabolism in oncology<br>• Assess myocardial hibernation<br>• Identify regions of abnormal glucose metabolism associated with foci of epileptic seizures |
| 7 | Fluorine-18 flutemetamol | GE Healthcare | Vizamyl | Indicated for PET imaging of the brain to estimate β amyloid neuritic plaque density in adult patients with cognitive impairment who are being evaluated for AD or other causes of cognitive decline |
| 8 | Gallium-67 citrate | Covidien | — | Useful to demonstrate the presence/extent of:<br>• Hodgkin's disease<br>• lymphoma<br>• Bronchogenic carcinoma |
| | | Lantheus Medical Imaging | — | Aid in detecting some acute inflammatory lesions |

*(Continued)*

**Table 6.2 (Continued)  Currently Approved Radiopharmaceuticals in the United States**

| # | Radiopharmaceutical | Manufacturer | Trade name | Indication |
|---|---|---|---|---|
| 9 | Indium-111 capromab pendetide | Jazz Pharmaceuticals | ProstaScint® | • A diagnostic imaging agent in newly-diagnosed patients with biopsy-proven prostate cancer, thought to be clinically-localized after standard diagnostic evaluation (e.g. chest x-ray, bone scan, CT scan, or MRI), who are at high risk for pelvic lymph node metastases<br>• A diagnostic imaging agent in post-prostatectomy patients with a rising PSA and a negative or equivocal standard metastatic evaluation in whom there is a high clinical suspicion of occult metastatic disease |
| 10 | Indium-111 chloride | Covidien | — | Indicated for radiolabeling:<br>• ProstaScint® used for in vivo diagnostic imaging procedures |
|  |  | GE Healthcare | Indiclor™ |  |
| 11 | Indium-111 pentetate | GE Healthcare | — | For use in radionuclide cisternography |
| 12 | Indium-111 oxyquinoline | GE Healthcare | — | Indicated for radiolabeling autologous leukocytes which may be used as an adjunct in the detection of inflammatory processes to which leukocytes migrate, such as those associated with abscesses or other infection |

*(Continued)*

**Table 6.2 (Continued)  Currently Approved Radiopharmaceuticals in the United States**

| 13 | Indium-111 pentetreotide | Octreoscan™ | Covidien | An agent for the scintigraphic localization of primary and metastatic neuroendocrine tumors bearing somatostatin receptors |
| 14 | Iodine 1-123 iobenguane | AdreView™ | GE Healthcare | **Indicated for use in the detection of primary or metastatic pheochromocytoma or neuroblastoma as an adjunct to other diagnostic tests Indicated for scintigraphic assessment of sympathetic innervation of the myocardium by measurement of the heart to mediastinum (H/M) ratio of radioactivity uptake in patients with New York Heart Association (NYHA) class II or class III heart failure and left ventricular ejection fraction (LVEF) $\leq$ 35%. Among these patients, it may be used to help identify patients with lower one and two year mortality risks, as indicated by an H/M ratio $\geq$ 1.6. Limitations of Use: In patients with congestive heart failure, its utility has not been established for: selecting a therapeutic intervention or for monitoring the response to therapy; using the H/M ratio to identify a patient with a high risk for death** |

*(Continued)*

**Table 6.2 (*Continued*)  Currently Approved Radiopharmaceuticals in the United States**

| | | | |
|---|---|---|---|
| 15 | Iodine 1-123 ioflupane[1] | GE Healthcare | DaTscan™ | Indicated for striatal dopamine transporter visualization using SPECT brain imaging to assist in the evaluation of adult patients with suspected Parkinsonian syndromes (PS) in whom it may help differentiate essential tremor due to PS (idiopathic Parkinson's disease, multiple system atrophy and progressive supranuclear palsy) |
| 16 | Iodine 1-123 sodium iodide capsules | **Cardinal Health** | — | **Indicated for use in the evaluation of thyroid:** <br>• **Function** <br>• **Morphology** |
| | | Covidien | — | |
| 17 | Iodine 1-125 human serum albumin | IsoTex Diagnostics | Jeanatope | **Indicated for use in the determination of:** <br>• **Total blood** <br>• **Plasma volume** |
| 18 | Iodine 1-125 iothalamate | IsoTex Diagnostics | Glofil-125 | Indicated for evaluation of glomerular filtration |

*(Continued)*

**Table 6.2 (*Continued*)   Currently Approved Radiopharmaceuticals in the United States**

| 19 | Iodine 1-131 human serum albumin | IsoTex Diagnostics | Megatope | Indicated for use in determinations of:<br>• **Total blood and plasma volumes**<br>• **Cardiac output**<br>• **Cardiac and pulmonary blood volumes and circulation times**<br>• **Protein turnover studies**<br>• **Heart and great vessel delineation**<br>• **Localization of the placenta**<br>• **Localization of cerebral neoplasms** |
| 20 | Iodine 1-131 sodium iodide | Covidien | — | Diagnostic:<br>• Performance of the radioactive iodide (RAI) uptake test to evaluate thyroid function<br>• Localizing metastases associated with thyroid malignancies<br>Therapeutic: |
| | | DRAXIMAGE | HICON™ | • Treatment of hyperthyroidism<br>• Treatment of carcinoma of the thyroid |

*(Continued)*

**Table 6.2 (*Continued*)   Currently Approved Radiopharmaceuticals in the United States**

| | | | |
|---|---|---|---|
| 21 | Molybdenum Mo-99 generator | Covidien | Ultra-TectineKow® DTE | Generation of $^{99m}$Tc sodium pertechnetate for administration or radiopharmaceutical preparation |
| | | GE Healthcare | DRYTEC™ | |
| | | Lantheus Medical Imaging | Technenlite® | |
| 22 | Nitrogen-13 ammonia[1] | Various | — | Indicated for diagnostic PET imaging of the myocardium under rest or pharmacologic stress conditions to evaluate myocardial perfusion in patients with suspected or existing coronary artery disease |
| 23 | Radium-223 dichloride | Bayer Healthcare Pharmaceuticals Inc | Xofigo® | Indicated for the treatment of patients with castration-resistant prostate cancer, symptomatic bone metastases and no known visceral metastatic disease |
| 24 | Rubidium-82 chloride | Bracco Diagnostics | Cardiogen-82® | PET myocardial perfusion agent that is useful in distinguishing normal from abnormal myocardium in patients with suspected myocardial infarction |

(*Continued*)

**Table 6.2 (Continued)   Currently Approved Radiopharmaceuticals in the United States**

| 25 | Samarium-153 lexidronam | Lantheus Medical Imaging | Quadramet® | Indicated for relief of pain in patients with confirmed osteoblastic metastatic bone lesions that enhance on radionuclide bone scan |
| 26 | Strontium-89 chloride | GE Healthcare | Metastron™ | Indicated for the relief of bone pain in patients with painful skeletal metastases that have been confirmed prior to therapy |
| 27 | Technetium-99m bicisate | Lantheus Medical Imaging | Neurolite® | SPECT imaging as an adjunct to conventional CT or MRI in the localization of stroke in patients in whom stroke has already been diagnosed |
| 28 | Technetium-99m disofenin | Pharmalucence | Hepatolite® | Diagnosis of acute cholecystitis as well as to rule out the occurrence of acute cholecystitis in suspected patients with right upper quadrant pain, fever, jaundice, right upper quadrant tenderness and mass or rebound tenderness, but not limited to these signs and symptoms |
| 29 | Technetium-99m exametazime | GE Healthcare | Ceretec™ | • As an adjunct in the detection of altered regional cerebral perfusion in stroke  • Leukocyte Labeled scintigraphy as an adjunct in the localization of intra-abdominal infection and inflammatory bowel disease |

*(Continued)*

**Table 6.2 *(Continued)*   Currently Approved Radiopharmaceuticals in the United States**

| | | | |
|---|---|---|---|
| 30 | Technetium-99m macroaggregated albumin | DRAXIMAGE | — | • An adjunct in the evaluation of pulmonary perfusion (adult and pediatric) • Evaluation of peritoneovenous (LaVeen) shunt patency |
| 31 | Technetium-99m mebrofenin | Bracco Diagnostics | Chotetec® | As a hepatobiliary imaging agent |
| | | Pharmalucence | — | |
| 32 | Technetium-99m medronate | Brae co Diagnostics | MDP Bracco® | As a bone imaging agent to delineate areas of altered osteogenesis |
| | | DRAXIMAGE | — | |
| | | DRAXIMAGE | MDP-25 | |
| | | GE Healthcare | MDP Multidose | |
| | | Pharmalucence | — | |

*(Continued)*

**Table 6.2 (Continued)   Currently Approved Radiopharmaceuticals in the United States**

| 33 | Technetium-99m mertiatide | Covidien | Technescan MAG3™ | In patients > 30 days of age as a renal imaging agent for use in the diagnosis of:<br>• Congenital and acquired abnormalities<br>• Renal failure<br>• Urinary tract obstruction and calculi<br>Diagnostic aid in providing:<br>• Renal function<br>• Split function<br>• Renal angiograms<br>• Renogram curves for whole kidney and renal cortex |
|---|---|---|---|---|
| 34 | Technetium-99m oxidronate | Covidien | Technescan™ HDP | As a bone imaging agent to delineate areas of altered osteogenesis (adult and pediatric use) |
| 35 | **Technetium-99m pentetate** | **DRAXIMAGE** | — | **• Brain imaging**<br>**• Kidney imaging:**<br> **• To assess renal perfusion**<br> **• To estimate glomerular filtration rate** |

*(Continued)*

**Table 6.2 (*Continued*)   Currently Approved Radiopharmaceuticals in the United States**

| | | | | |
|---|---|---|---|---|
| 36 | Tech return-99m pyrophosphate | Covidien | Tech rescan™ PYP™ | • As a bone imaging agent to delineate areas of altered osteogenesis<br>• As a cardiac imaging agent used as an adjunct in the diagnosis of acute myocardial infarction<br>• As a blood pool imaging agent useful for:<br>  • Gated blood pod imaging<br>  • Detection of sites of gastrointestinal bleeding |
| | | Pharmalucence | — | |
| 37 | Technetium-99m red blood ceils | Covidien | UltraTag™ | Tc99m-labeled red blood cells are used for:<br>• Blood pool imaging including cardiac first pass and gated equilibrium imaging<br>• Detection of sites of gastrointestinal bleeding |

*(Continued)*

Table 6.2 (*Continued*)  Currently Approved Radiopharmaceuticals in the United States

| 38 | Technetium-99m sestamibi | Cardinal Health | — | Myocardial perfusion agent that is indicated for: <br>• Detecting coronary artery disease by localizing myocardial ischemia (reversible defects) and infarction (non-reversible defects) <br>• Evaluating myocardial function <br>• Developing information for use in patient management decisions <br><br>Planar breast imaging as a second line diagnostic drug after mammography to assist in the evaluation of breast lesions in patients with an abnormal mammogram or a palpable breast mass |
| | | Covidien | — | |
| | | DRAXIMAGE | — | |
| | | Lantheus Medical Imaging | Cardiolite® | |
| | | Pharmalucence | — | |
| 39 | Technetium-99m sodium pertechnetate | Covidien | — | • Brain Imaging (including cerebral radionuclide angiography)* <br>• Thyroid Imaging* <br>• Salivary Gland Imaging <br>• Placenta Localization <br>• Blood Pool Imaging (inducing radionuclide angiography)* <br>• Urinary Bladder Imaging (direct isotopic cystography) for the detection of vesicoureteral reflux* <br>• Nasolacrimal Drainage System Imaging (*adult and pediatric use) |
| | | GE Healthcare | — | |
| | | Lantheus Medical Imaging | — | |

(*Continued*)

**Table 6.2 (*Continued*)  Currently Approved Radiopharmaceuticals in the United States**

| 40 | Technetium-99m succimer | GE Healthcare | — | An aid in the scintigraphic evaluation of renal parenchymal disorders |
|----|------------------------|---------------|---|-----------------------------------------------------------------------|
| 41 | Technetium-99m sulfur colloid | Pharmalucence | — | • Imaging areas of functioning reticuloendothelial cells in the liver, spleen and bone marrow* <br> • It is used orally for: <br>   • Esophageal transit studies* <br>   • Gastroesophageal reflux scintigraphy* <br>   • Detection of pulmonary aspiration of gastric contents* <br> • Aid in the evaluation of peritoneovenous (LeVeen) shunt patency <br> • To assist in the localization of lymph nodes draining a primary tumor in patients with breast cancer or malignant melanoma when used with a hand held gamma counter (*adult and pediatric use) |

*(Continued)*

**Table 6.2 (Continued)  Currently Approved Radiopharmaceuticals in the United States**

| 42 | Technetium-99m tetrofosmin | GE Healthcare | Myoview™ | Myocardial perfusion agent that is indicated for: • Detecting coronary artery disease by localizing myocardial ischemia (reversible defects) and infarction (non-reversible defects) • The assessment of left ventricular function (left ventricular ejection fraction and wall motion) |
| --- | --- | --- | --- | --- |
| 43 | Technetium-99m tilmanocept | Navidea Biopharmaceuticals, Inc. | Lymphoseek® | Indicated with or without scintigraphic imaging for: • Lymphatic mapping using a handheld gamma counter to locate lymph nodes draining a primary tumor site in patients with solid tumors for which this procedure is a component of intraoperative management • Guiding sentinel lymph node biopsy using a handheld gamma counter in patients with clinically node negative squamous cell carcinoma of the oral cavity, breast cancer or melanoma |

*(Continued)*

**Table 6.2 (Continued)   Currently Approved Radiopharmaceuticals in the United States**

| 44 | Thallium-201 chloride | Covidien | — | • Useful in myocardial perfusion imaging for the diagnosis and localization of myocardial infarction |
|----|----|----|----|----|
| | | GE Healthcare | — | • As an adjunct in the diagnosis of ischemic heart disease (atherosclerotic coronary artery disease) |
| | | Lantheus Medical Imaging | — | • Localization of sites of parathyroid hyperactivity in patients with elevated serum calcium and parathyroid hormone levels |
| 45 | Xenon-133 gas | Lantheus Medical Imaging | — | • The evaluation of pulmonary function and for imaging the lungs<br>• Assessment of cerebral flow |
| 46 | Yttrium-90 chloride | MDS Nordion | — | **Indicated for radiolabeling:** |
| | | Eckert & Ziegler Nuditec | — | • Zevalin® used for radioimmunotherapy procedures |
| 47 | Yttrium-90 ibritumomab tiuxetan | Spectrum Pharmaceuticals | Zevalin® | Indicated for the:<br>• Treatment of relapsed or refractory, low-grade or follicular B-cell non-Hodgkin's lymphoma (NHL)<br>• Treatment of previously untreated follicular NHL in patients who achieve a partial or complete response to first-line chemotherapy |

*Source:* Courtesy of Cardinal Health, Dublin, OH.

## Chapter 7

# Current Pharmaceuticals Used in Nuclear Medicine Therapy

A variety of radiopharmaceuticals are used in nuclear medicine therapy. The use of radioiodines to treat thyroid diseases and $^{32}$P to treat polycythemia vera (Ferreira, Rosario Vieira, and de Jonge 2007) (which is no longer in use) was established decades ago. Development and investigation of new agents is always progressing; an important issue, however, is clinical acceptance of new therapies that are intended to replace existing therapies. Resistance to change can cause difficulties in sustaining new products, as the approval process for new drugs is quite expensive, and poor market performance has caused distribution of some very good agents to be discontinued. Nonetheless, some very effective new agents have been developed recently, and the future of radiopharmaceutical therapy is bright. In this chapter, we will overview the existing agents and their application. In Chapter 8, we will review how implementation of patient-individualized dosimetry for these therapies is needed to optimize the effectiveness of these agents; at present, this is not a common practice.

# Classes of Pharmaceuticals

## Thyroid Diseases

Radioactive iodine ($^{131}$I) is used in the treatment of hyper-thyroidism and thyroid cancer, usually using a fixed-dosage approach. Oliveira and Pereira de Oliveira (2007) note: "Although the clinical merits of dosimetry-guided radioiodine therapy have been demonstrated, most centers have adapted the fixed-dose technique using 3.7 to 7.4 GBq (100 to 200 mCi) I-131 owing to the technical and logistic difficulties of dosimetry studies." Maxon et al. (1983) showed that radioiodine uptake and clearance half-times varied substantially and that positive outcomes were clearly associated with achieving a certain radiation dose, calculated individually for each patient. Uptake in thyroid remnants varied from 0.1 to 27% (mean 7.4%), and half-times varied from 19 h to 180 h (mean 103 h) in patients whose ablations were successful. On the contrary, patients with unsuccessful ablations had uptakes in thyroid remnants that varied from 0.3 to 30% (mean 10.2%), and half-times varied from 15 h to 156 h (mean 56 h). The dose ranges were 128 Gy to 7,000 Gy (mean 1,150 Gy) for the patients with good outcomes, but only 52 Gy to 885 Gy (mean 264 Gy) in the patients with poor outcomes. The enormous *technical and logistic difficulties* involved taking measurements of activity in the neck region at 24, 48, and 72 h after the diagnostic (74 MBq) administration of radioiodine. Compared to the effort invested in every external beam therapy patient (discussed in the next chapter), this is a very small amount of work that will clearly result in better outcomes for many patients; the only issue is that patients need to return to be imaged on separate days. If you ask patients if they want certainty of a positive outcome from their cancer, it is difficult to imagine many refusing this simple inconvenience. In a later study with 85 patients, Maxon et al. (1992) showed that a dosimetry-based approach gave high rates of success (around 80%) for treatment of thyroid remnants if doses were above 300 Gy, and lymph node metastases were successfully treated in 74% of patients if doses were above 85 Gy. They concluded: "The quantitative radiation dosimetric approach to [131]I treatment results in predictably high success rates while allowing the assignment of higher exposures and of greater costs to those individuals who require them." Kobe et al. (2008) evaluated the success of treatment of Graves' disease in 571 subjects, with the goal of delivering 250 Gy to the thyroid, with the end-point measure being the elimination of hyperthyroidism, evaluated 12 months after the treatment. Relief from hyperthyroidism was achieved in 96% of patients who received more than 200 Gy, even for those with thyroid volumes greater than 40 ml (Figure 7.1).

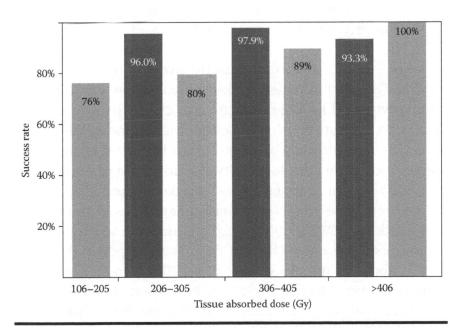

**Figure 7.1   Three-month success rates in treating Graves' disease when individual patient thyroid doses are evaluated and kept above 200 Gy. (From Kobe C et al., 2008, *Nuklearmedizin*, 47, 13–17.)**

## Lymphomas

Diffuse large B-cell lymphoma (DLBCL) and follicular lymphoma may be treated with chemotherapeutic regimens, with median durations of about 10 years. Morgan, Wardy, and Barton (2004) give the 5-year survival to be about 10% for non-Hodgkin's lymphoma (NHL). Monoclonal antibodies, targeting the CD20 antigen, can cause the stimulation of plasma cells to produce immunoglobulins that may produce antitumor effects, depending on the degree of binding to tumor cells. One immunoglobulin, ibritumomab, is a chimeric antibody modified to minimize the development of human anti-murine antibodies (HAMA) in patients. The antibody was administered in conjunction with a chemotherapy regime; unfortunately, over time, many patients were found to be refractory to these infusions. Forms of the antibody labeled with either $^{131}$I or $^{90}$Y were developed, in the hope that radiation effects on the tumor, combined with the antibody effects, would be more effective.

Several studies evaluated the effectiveness of the $^{131}$I product (Bexxar) in patients with follicular, low-grade lymphoma who had become refractory to chemotherapy and rituximab. A relatively simple dosimetric regime is followed, as response has been shown to be better related to whole-body dose than to activity administered on a per-unit body weight or surface area approach; variability in patients' effective whole-body half-times varies as much as a factor of 5 (Wahl 2005). Phase 3 trials demonstrated overall response rates (ORR) of 56%, with response duration of about 13 months (Fisher et al. 2005), and a complete response (CR) rate of 30%, with response duration of 28 months. In patients refractory to rituximab infusions, an ORR of 65% and a CR of 29% were found; the duration of response was 35 months. Follicular grade 1 or 2 patients with tumors smaller than 7 cm had OR and CR rates of 86% and 57%, respectively (Horning et al. 2005). These are very good results, but these patients have been through several cycles of chemotherapy and immunotherapy. When Bexxar was used as part of an initial regime after administration of fludarabine, 76 patients with stage 3 or 4 follicular lymphoma had OR and CR rates of 97% and 76%, respectively, and a median progression-free survival (PFS) of 48 months and low HAMA rates (6%, Goldsmith 2013). Another multicenter trial found an ORR of 96% in 76 patients with NHL, a CR of 76%; the mean time to progression (TTP) was not observed in 6 years of patient follow-up. Immunosuppression was observed, but modest.

Clinical trials with the $^{90}$Y antibody (Zevalin) involve pretreatment with rituximab 1 week before therapy with Zevalin. Activity is administered based on patient body weight and patient platelet count, with no treatment of dosimetry. Myelosuppression is avoided by using a cautious approach, and, as noted in the preceding text, this implies that many patients are not receiving as much activity as they could likely tolerate, as nothing is known about their whole-body clearance kinetics. Initial trials showed an ORR of 80% and a CR of 30%, with response duration of 14 months. Myelosuppression was observed, but again was manageable and reversible. A following study of 57 patients not responding to rituximab therapy had an ORR of 74% and a CR of 15%, with a TTP of 8.7 months for patients with any response.

With these excellent response rates, one would imagine that these drugs would be widely used. Their use, however, was far less than expected, and Bexxar recently has been removed from the market. Goldsmith notes the reluctance of physicians to employ radioimmunotherapy due to concerns about bone marrow suppression causing hematalogic consequences and possibly interfering with future chemotherapy regimes, despite evidence that suppression is generally mild and reversible. He goes on to note: "Despite the considerable successes to date, many patients have been denied access to these regimen as well as the newer experimental modalities because of lack of enthusiasm which is a result of the lack of understanding of the features, safety and merits of radioimmunotherapy." Press also notes *market-driven forces* that cause physicians to choose other, less effective methods that are more difficult for the patients (e.g., chemotherapy); administration needs to be performed in a medical center, not doctors' offices, and thus many doctors are not paid by Medicare and private insurers when they choose to use the radioactive antibodies (Berenson 2007). Press says: "Both Zevalin and Bexxar are very good products. It's astounding and disappointing [that they are used so little]."

## *Neuroendocrine Tumors*

Neuroendocrine tumors (NETs) are a class of tumors that expresses somatostatin receptors (SSTRs). Peptide-receptor radionuclide therapy (PRRT) uses somatostatin analogs to image and treat NETs. Examples include $^{111}$In-DTPA-octreotide, $^{90}$Y-DOTATOC, $^{177}$Lu-DOTATATE, $^{90}$Y-DOTATATE, and $^{131}$I metaiodobenzylguanidine (MIBG). PRRT may result in significant doses to bone marrow and kidney. Two studies have shown that patient variability in biokinetics and renal excretion vary markedly between patients, and that a regime that includes calculation of radiation dose is essential to managing the radiotoxicity (Barone et al. 2005; Bodei et al. 2008). These authors found, however, that typical calculations of absorbed dose (Gy, J/kg) were not sufficient to explain the kidney response to absorbed dose, and that a parameter called the biologically effective dose (BED), which takes into account the rate at which the dose is received and the theoretical parameters describing tissue repair rates, provided a better explanation. Siegel, Stabin, and Sharkey (2010) showed that another approach using *time–dose fractionation* (TDF) factors provided better correlations of changes in serum creatinine levels than BED; they also discussed complications in the calculation of the appropriate calculation of kidney dose and use of radiobiological models and concluded that this issue should be explored with more data and model approaches. Cremonesi et al. (2006), in a review of the literature regarding PRRT, noted the significant variability of patients' organs and, more notably, tumor uptake and clearance, and stressed the need for a personalized approach to the choice of radiopharmaceutical(s) for each patient, and patient-individualized evaluation of radiation dose and response. Current multicenter trials in the United States, however, are based on a fixed-dosing scheme based only on patient body weight; this is also true in Europe, with only a few exceptions. If maintained at a low and safe level, renal toxicity may be avoided, but, again, many patients may be getting suboptimal therapy, and there is no way to relate renal toxicity to dose and dose-rate-based parameters when it will be observed.

Dose to bone marrow is limiting for $^{131}$I-MIBG. Overall response rates are good (35%), and are similar to results for chemotherapy, with significantly less toxicity and discomfort for the patients. Many studies have found that doses to tumors and dose-limiting normal tissues can vary by an order of magnitude, and that personalized dosimetry is essential to optimizing each patient's therapy while maintaining marrow doses at a safe level (Flux et al. 2011). Estimating, at a minimum, whole-body clearance using three to four whole-body scans is essential; some have even used a mounted radiation probe instead of a gamma camera to implement this simple approach to whole-body/marrow dosimetry. Whole-body doses are notably different for children with neuroblastoma (0.37 + 0.21 mGy/MBq) than for adults with pheochromocytomas (0.08 + 0.02 mGy/MBq) (Monsieurs et al. 2002). Performing tumor and organ dosimetry requires collection and quantification of image data. Flux et al. (2011) note, however, that this is logistically feasible, and values compared between two centers showed good reproducibility of results; target doses were delivered within 10% of the desired values in 90% of the cases. Statistically significant correlations were found between calculated whole-body doses and neutropenia, but not between administered activities and hematologic toxicity (Buckley et al. 2009). Flux et al. conclude: "The wide variation in delivered absorbed doses and in observed responses indicate that personalized dosimetry is essential to optimize this therapy, and there is now sufficient evidence to demonstrate that this is feasible." Yet, most centers continue to treat with a fixed-dosing approach, providing poor therapy for the patients and unnecessarily risking the patients developing complications from the marrow doses received.

## Hepatic Tumors

Hepatic tumors can be treated via injection of microspheres labeled with $^{90}Y$ directly into the hepatic artery. The microspheres are taken up in higher concentrations in tumor than in normal tissue, due to differences in blood flow. A therapeutically effective dose can usually be delivered to tumor with limited radiation dose and deleterious effects in the normal liver tissue. Still, attention has to be given to the tumor-to-liver concentrations in a given patient, and there is the chance of unwanted *shunting* of the microspheres to lung. This shunting is evaluated via administration of $^{99m}Tc$ macroaggregated albumin (MAA) particles; there is some controversy regarding how predictive this method is. Chiesa et al. (2015) found that biodistributions based on $^{99m}Tc$ MAA and $^{90}Y$ bremsstrahlung imaging were not different in 71% of cases, different in 23%, and uncertain in 6%. As the microspheres are trapped in the capillaries of the various tissues, a single image can produce reliable dosimetry, as the effective half-life is just the physical half-life of $^{90}$Y. The dosimetry is also very simple, as there is only a single beta emission that provides the therapeutic effect. With a CT image of the liver and tumors to provide volumes, doses to liver, tumor, and the percentage of lung involvement are calculated with very simple equations, which is much easier than whole-body and organ/tumor doses when activity is distributed widely in the body. Dosimetry for this agent can be performed with simple spreadsheet calculations once the few pieces of information noted in the preceding text are gathered. Performing therapy using a fixed-dosing approach is really indefensible for these agents.

Threshold tumor doses for good responses are different between resin microspheres (80–100 Gy) and glass microspheres (perhaps 250 Gy) (Garin et al. 2013; Chiesa et al. 2015). Several trials have shown clear improvements in ORR and 5-year survival when chemotherapy was augmented with therapy with radioactive microspheres in patients with hepatic colorectal metastases (Gulec, Suthar, and Barot 2013). Lau et al. (1994) reported partial responses of 67% and complete responses of 22% in patients treated for hepatocellular carcinoma, and treatment efficacy was correlated with the radiation dose delivered. PFS was significantly different in patients whose tumors received over 120 Gy; median survival in patients with doses below 120 Gy was only 26 weeks, whereas those with doses above 120 Gy had PFS of 56 weeks. Chiesa et al. (2015) tested the effectiveness of radiobiological dose quantities equivalent uniform dose (EUD) and equivalent uniform biologically effective dose (EUBED) (see Chapter 1), and found, for normal tissue complication probabilities (NTCPs), that no method was superior to parenchyma mean dose, as a dose rate effect was not seen (Figure 7.2). They asserted that this quantity was adequate for therapy planning with a 15% tolerance dose of about 75 Gy.

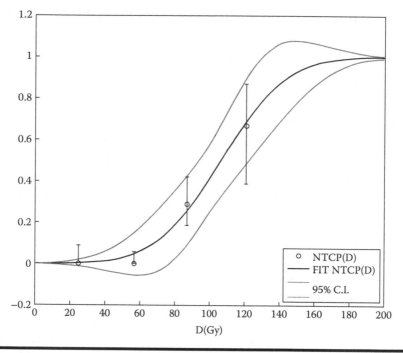

**Figure 7.2 Normal tissue complication probability (NTCP) as a function of the mean absorbed dose averaged over the parenchyma, excluding tumor. (From Chiesa C et al., 2015, *Eur J Nucl Med Mol Imaging*, 42, 1718–1738.)**

## Bone-Seeking Agents

Bone-seeking radiopharmaceuticals are employed to treat the pain of skeletal metastases, typically from breast or prostate cancer. Hydroxyapatite is the calcium phosphate complex $(Ca_{10}(PO_4)_6(OH)_2)$ (Figure 7.3) that is the primary mineral component of the skeleton (and teeth). Substances that mimic either the metabolic behavior of calcium, phosphate, or the OH ion will have affinity for the skeleton and may be useful in treating these painful bony metastases. This kind of therapy has been in practice for many decades. Until recently, all clinically employed agents were beta emitters; recently, the alpha emitter $^{223}$Ra has been approved for clinical use and has shown good effects in extending the life of patients with bony metastases.

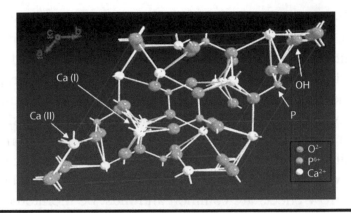

**Figure 7.3   Structure of hydroxyapatite. (From Medical MCQs, http://www.medicalmcqs.com/wp-content/uploads/2013/12/ Hydroxyapatite.png.)**

$^{223}$Ra has a series of radioactive progeny that deliver a radiation dose from several alpha and beta emissions. In general, the aim of this kind of therapy is only palliative, not curative.

The effectiveness of an agent will depend on two key quantities: (1) the range of the principal emission and (2) the physical half-life of the radionuclide. Lewington (2005) provided a summary of the physical characteristics of some important bone-seeking radiopharmaceuticals.

| Radionuclide | Half-life | Maximum Energy (MeV) | Mean Energy (MeV) | Maximum Range | γ Emission (keV) |
|---|---|---|---|---|---|
| $^{32}$P | 14.3 d | 1.7 (β) | 0.695 (β) | 8.5 mm | None |
| $^{89}$Sr | 50.5 d | 1.4 (β) | 0.583 (β) | 7 mm | None |
| $^{186}$Re | 3.7 d | 1.07 (β) | 0.362 (β) | 5 mm | 137 |
| $^{188}$Re | 16.9 h | 2.1 (β) | 0.764 (β) | 10 mm | 155 |
| $^{153}$Sm | 1.9 d | 0.81 (β) | 0.229 (β) | 4 mm | 103 |
| $^{117m}$Sn | 13.6 d | 0.13, 0.16 CE[a] | | <1 | 159 |
| $^{223}$Ra | 11.4 d | 5.78 (α) | | <10 | 154 |

[a] Conversion electrons.

- *$^{32}$P-phosphate*—This agent has been used for many years; it binds to hydroxyapatite as inorganic phosphate. Pain relief has been observed in 50–90% of treated patients, with typical activities of 200–800 MBq. Response times are of the order of 2 weeks, and response duration may be as much as 10 weeks. Its 1.7 MeV maximum beta energy results in significant doses to bone marrow, and myelosuppression is the dose-limiting effect. Lewington (2005) notes that toxic deaths are rare, but marrow effects are common and must be monitored carefully. The agent is not used much in the United States and Europe, but is employed elsewhere due to its low cost and ease of administration.

- *$^{89}$Sr-chloride*—Strontium is a calcium analogue; uptake in bone is good, and tumor to normal bone uptake ratios are around 10:1. Tumor doses over 2 Gy/MBq have been calculated, although correlations of dose and effect have not been established. Pain relief is seen in over 70% of patients (McEwan 1998). Toxicity issues are also related to myelosuppression, perhaps over up to 6 weeks post-therapy.

- *$^{186}$Re and $^{188}$Re hydroxyethylidene diphosphonate (HEDP)*—HEDP is a phosphate-type molecule that forms stable complexes with these rhenium isotopes. $^{188}$Re has a high beta energy (2.1 MeV maximum), while that of $^{186}$Re is more moderate (1.07 MeV). Skeletal uptake is proportional to tumor burden, and excretion is principally renal. Beneficial responses have been reported for 80–90% of $^{186}$Re patients and 75% of $^{188}$Re patients. Toxicity is again related to myelosuppression, reaching WHO grades of 3–4 in studies with $^{188}$Re (Palmedo et al. 2000).

- *$^{153}$Sm-ethylenediaminetetramethylenephosphonate (EDTMP)*—EDTMP is similar to HEDP—a phosphonate compound that mimics phosphate. It also forms stable complexes with samarium. Skeletal uptake ranges from 55% to 75%, again in relation to tumor burden. Pain relief has been reported in 65–80% of patients, with a fairly rapid response. Myelosuppression is again the limiting toxicity; marrow dose should be limited to 2 Gy.

■ *$^{117m}$Sn-diethylenetriaminepentaacetic acid (DTPA)*—$^{117m}$Sn emits very short-range conversion electrons. Chelated with DTPA, it demonstrates over 75% uptake in bone, with no observable biologic removal. Pain relief has been reported in 75–90% of patients in clinical trials, with minimal myelotoxicity.

■ *$^{223}$Ra-chloride*—As with strontium, radium is a calcium analogue. It is relatively short lived, and has uptake on the surfaces of growing bone. Unlike the other agents discussed in the preceding text, it has significant clearance through the gastrointestinal tract, and some manageable effects (diarrhea, nausea, vomiting) have been reported. The principal toxicity is again myelosuppression, but at low grades. Response rates have been good, and a statistically significant difference in life extension was seen (14.9 vs. 11.3 months) over placebos.

# Bibliography

Barone R, Borson-Chazot F, Valkema R, Walrand S, Chauvin F, Gogou L, Kvols LK, Krenning EP, Jamar F, and Pauwels S. 2005. Patient-specific dosimetry in predicting renal toxicity with $^{90}$Y-DOTATOC: relevance of kidney volume and dose rate in finding a dose-effect relationship. *J Nucl Med*, 46(suppl), 99S.

Berenson A. 2007. Market forces cited in lymphoma drugs' disuse. *New York Times*, July 31.

Bodei L, Cremonesi M, Ferrari M, Pacifici M, Grana CM, Bartolomei M, Baio SM, Sansovini M, and Paganelli G. 2008. Long-term evaluation of renal toxicity after peptide receptor radionuclide therapy with $^{90}$Y-DOTATOC and $^{177}$Lu-DOTATATE: the role of associated risk factors. *Eur J Nucl Med Mol Imaging*, 35, 1847.

Buckley SE, Chittenden SJ, Saran FH, Meller ST, and Flux GD. 2009. Whole-body dosimetry for individualized treatment planning of $^{131}$I-MIBG radionuclide therapy for neuroblastoma. *J Nucl Med*, 50(9), 1518–1524.

Chiesa C, Mira M, Maccauro M, Spreafico C, Romito R, Morosi C, Camerini T, et al. 2015. Radioembolization of hepatocarcinoma with $^{90}$Y glass microspheres: development of an individualized treatment planning strategy based on dosimetry and radiobiology. *Eur J Nucl Med Mol Imaging*, 42(11), 1718–1738.

Cremonesi M, Ferrari M, Bodei L, Tosi1 G, and Paganelli G. 2006. Dosimetry in peptide receptor therapy: a review. *J Nucl Med*, 47, 1467.

Ferreira TC, Rosario Vieira M, and de Jonge FAA. 2007. Polycythemia Vera and Therapy with Phosphorus-32. In *Nuclear Medicine Therapy*, edited by J Eary and W Brenner. Informa Healthcare, New York.

Fisher RI, Kaminski MS, Wahl RL, Knox SJ, Zelenetz AD, Vose JM, Leonard JP, Kroll S, Goldsmith SJ, and Coleman M. 2005. Tositumomab and iodine-131 tositumomab produces durable complete remissions in a subset of heavily pretreated patients with low-grade and transformed non-Hodgkin's lymphomas. *J Clin Oncol*, 23, 7565–7573.

Flux GD, Chttenden SJ, Saran F, and Gaze MN. 2011. Clinical applications of dosimetry for mIBG therapy. *Q J Nucl Med Mol Imaging*, 55, 116–125.

Garin E, Lenoir L, Edeline J, Laffont S, Mesbah H, Poree P, Sulpice L, et al. 2013. Boosted selective internal radiation therapy with $^{90}$Y-loaded glass microspheres (B-SIRT) for hepatocellular carcinoma patients: a new personalized promising concept. *Eur J Nucl Med Mol Imaging*, 40(7), 1057–1068.

Goldsmith SJ. 2013. Radionuclide therapy of lymphomas. In *Nuclear Medicine Therapy, Principles and Clinical Applications*, edited by C Aktolun and SJ Goldsmith. Springer, New York, NY.

Gulec SA, Suthar R, and Barot T. 2013. Radiomicrosphere therapy of liver tumors. In *Nuclear Medicine Therapy, Principles and Clinical Applications*, edited by C Aktolun and SJ Goldsmith. Springer, New York, NY.

Horning SJ, Younes A, Jain V, Kroll S, Lucas J, Podoloff D, and Goris M. 2005. Efficacy and safety of tositumomab and iodine-131 tositumomab (Bexxar) in B-cell lymphoma, progressive after rituximab. *J Clin Oncol*, 23, 712–719.

Kobe C, Eschner W, Sudbrock F, Weber I, Marx K, Dietlein M, and Schicha H. 2008. Graves' disease and radioiodine therapy: is success of ablation dependent on the achieved dose above 200 Gy? *Nuklearmedizin*, 47(1), 13–17.

Lau WY, Leung WT, Ho S, Leung NWY, Chan M, Lin J, Metreweli C, Johnson P, and Li AKC. 1994. Treatment of inoperable hepatocellular carcinoma with intrahepatic arterial yttrium-90 microspheres: a phase I and II study. *Br J Cancer*, 70, 994–999.

Lewington VJ. 2005. Bone seeking radionuclides for therapy. *J Nucl Med*, 46(1), 38S–47S.

Maxon HR, Thomas SR, Hertzberg VS, Kereiakes JG, Chen IW, Sperling MI, and Saenger EL. 1983. Relation between effective radiation dose and outcome of radioiodine therapy for thyroid cancer. *N Engl J Med*, 309(16), 937–941.

Maxon HR, Englaro EE, Thomas SR, Hertzberg VS, Hinnefeld JD, Chen LS, Smith H, Cummings D, and Aden MD. 1992. Radioiodine-131 therapy for well-differentiated thyroid cancer— a quantitative radiation dosimetric approach: outcome and validation in 85 patients. *J Nucl Med*, 33, 1132–1136.

McEwan AJB. 1998. Palliation of bone pain. In *Nuclear Medicine in Clinical Diagnosis and Treatment*, edited by IPC Murray and PJ Ell (pp. 1083–1099). Churchill Livingstone, London, UK.

Monsieurs M, Brans B, Bacher K, Dierckx R, and Thierens H. 2002. Patient dosimetry for [131]I-MIBG therapy for neuroendocrine tumors based on [123]I-MIBG scans. *Eur J Nucl Med Mol Imaging*, 29(12), 1581–1587.

Morgan G, Wardy R, and Barton M. 2004. The contribution of cyto-toxic chemotherapy to 5-year survival in adult malignancies. *Clin Oncol*, 16, 549–560.

Oliveira MJ and Pereira de Oliveira. 2007. Treatment of differ-entiated thyroid carcinoma. In *Nuclear Medicine Therapy*, edited by J Eary and W Brenner. Informa Healthcare, New York, NY.

Palmedo H, Guhlke S, Bender H, Sartor J, Schoeneich G, Risse J, Grünwald F, Knapp FF Jr, and Biersack HJ. 2000. Dose escala-tion study with rhenium-188 hydroxyethylidene diphosphonate in prostate cancer patients with osseous metastases. *Eur J Nucl Med*, 27, 123–130.

Siegel JA, Stabin MG, and Sharkey RM. 2010. Renal dosimetry in peptide radionuclide receptor therapy. *Cancer Biother Radio*, 25(5), 581–588.

Wahl RL. 2005. Tositumomab and [131]I therapy in non-Hodgkin's lymphoma. *J Nucl Med*, 46, 128S–140S.

## Chapter 8

# The Need for Patient-Individualized Dosimetry in Therapy

Patients receiving therapy with radiopharmaceuticals currently are receiving poor-quality medical care. Instead of tailoring the therapy to each individual patient, based on his or her specific biochemistry and metabolism, physicians generally prescribe dosages of radiopharmaceuticals to treat cancer and other diseases in a *one-size-fits-all* approach. Some administrations are adjusted for patient body weight, but this does not account for each subject's specific uptake of the compounds in tumors and normal tissues, and this may be inappropriate for obese patients, as organ sizes and radionuclide uptakes and clearance rates in their organs and tissues may be the same as a normal-weight individual. In external beam radiotherapy, every single patient receives an individualized radiation dose evaluation before therapy begins, based on computed tomography (CT) images of the subject and testing of computer-simulated therapy treatments. Thus, patients receive individualized treatment plans, which have the best chance of success with their disease. Giving 100 nuclear medicine therapy patients the same activity to treat their disease results in a broad distribution of effectiveness in the therapy and the outcomes, as much as a factor of 5 or more, as measured for one nuclear medicine therapy agent (Wahl 2005). As this is known, the tendency is to err on the side of caution and give too little, rather than too much radiation, to avoid normal tissue complications. Hence, if the distribution of effectiveness can be assumed to be Gaussian, the average of the distribution will be suboptimal, a large proportion of the population receives therapy that is decidedly suboptimal, and only a fraction of this population can expect a favorable therapeutic effect.

The goal of any cancer therapy is to eradicate malignant tissues, or at least arrest their growth, while limiting toxicity to manageable levels in normal tissues. There are many agents that make this possible. Radioactive iodine is avidly taken up by the thyroid and thyroid cancer metastases in most patients, with minimal or no adverse effects on the red marrow. The science of targeted therapy, for example, with monoclonal antibodies, has been developed to a mature level, with many agents and strategies for optimizing their therapeutic effects against several types of cancer. These agents, in general, operate with a much narrower *therapeutic window*, however, and, to have a good chance of success, therapy for each patient should be individually tailored based on dosimetry established with a tracer level of activity. Radiation doses will vary significantly among different patients, and, furthermore, different patients' responses to a given radiation dose will not be the same, due to their individual characteristics and pre-existing conditions (e.g., effects of prior chemotherapy, complications such as poor renal function). Much remains to be learned about how to appropriately apply patient-individualized dosimetry to therapy planning; one of the arguments that opponents make is that we do not have sufficient knowledge at present to use dosimetry effectively. Hence, fixed-dosing approaches continue, with patients receiving a poor quality of therapy in many cases;

and, with no knowledge of what radiation doses they may have received, we can never advance this knowledge.

We measure or calculate radiation doses for radiation workers receiving low doses at low dose rates (this is required by law, of course), for patients receiving diagnostic levels of radiation dose from CT scans and other studies, for anyone receiving radiation therapy using external sources (radiotherapy, brachytherapy), and even for airline pilots and flight attendants. The only people on earth for whom we do not calculate radiation doses are nuclear medicine patients. These patients, many of them children, are treated with fixed-dosing techniques; no retrospective dosimetry is performed; and, years later, if we need to know how much dose they received from these studies, for cancer epidemiology studies or further therapies, we have absolutely no idea of what radiation doses they may have received. *Patient-individualized medicine* is a current area of growth, tailoring patients' treatments and pharmaceutical choices to their individual makeup and genetic history. Only in nuclear medicine therapy do we pay no attention to individual patients' characteristics and needs and treat them with substandard medical care.

Methods and tools are well defined for performing patient-specific dosimetry in radiopharmaceutical therapy. In this chapter, we will discuss several forms of radiopharmaceutical therapy, and discuss how radiation dosimetry methods have shown effectiveness in optimizing patient therapy (i.e., maximizing dose to tumors while managing toxicity in normal tissues), with comparison to rates of success using chemotherapy.

## Comparison to External Beam Therapy

Chemotherapeutic drugs, whether given with the intent to cure or just prolong life and palliate negative symptoms, are typically given based on a subject's body surface area (BSA).

Again, for obese patients, this may result in prescribing a very high dosage that may not be safe, so upper limits are usually established for any administration. But there is no accounting for each individual's clearance kinetics, and persons with similar body size may have significantly different rates of clearance of the drug, and thus significantly different possibilities of effective therapy and significantly different levels of side effects. Calvert et al. (1989) showed how to use individual subjects' biokinetics, specifically the area under concentration/time curves (AUC) for the agent carboplatin, to tailor dosages to patients with ovarian, lung, testicular, head and neck, and pancreatic cancer and mesothelioma. Starting with a low-level, fixed-dosage administration, the patient's glomerular filtration rate (GFR) was measured, which has been shown to correlate well with the AUC. Then, dosages were escalated using the patient-specific parameters, and measured and predicted AUCs were well correlated. Newell et al. (1993) extended the method to pediatric subjects; they noted "a wide range of values for absolute and dose-normalized carboplatin AUC, clearance, half-life (t1/2), and volume of distribution." For example, t1/2 values had a mean of 19 minutes, with a range of 3–52 minutes. Calvert et al. conclude that "The [GFR-based] formula compensates for variations in pretreatment renal function that might otherwise result in either underdosing (in patients with above average renal function) or overdosing (in patients with impaired renal function). As carboplatin plasma AUC, rather than toxicity, is the measure of drug exposure defined by the formula, it can be applied to both combination and high-dose therapy. It is recommended that future studies with carboplatin be conducted with the dose determined by the formula described here to ensure that the full therapeutic potential of this drug is realized."

Five-year survival for all chemotherapy agents is only 2.3% on average (Morgan et al. 2004), with values between 0.7% and 36% (the latter being Hodgkin's disease). Hoskins et al. (2001) reported a 5-year overall survival in 41 patients treated with fixed-dose paclitaxel and patient-individualized carboplatin regime (according to the Calvert method) of over 60%. In two prospectively randomized, controlled studies conducted by the National Cancer Institute of Canada, Clinical Trials Group (NCIC) and the Southwest Oncology Group (SWOG) involving patients with ovarian cancer, 3-year progression-free survival (PFS) rates were 19% and 8%, respectively, and 3-year survival rates were 34.6% and 18.3%, respectively.

# Common Reasons for Resistance to Patient-Individualized Dosimetry and Arguments for Overcoming the Resistance

As Siegel, Stabin, and Brill noted (2002):

> If one were to approach the radiation oncologist or medical physicist in an external beam therapy program and suggest that all patients with a certain type of cancer should receive the exact same protocol (beam type, energy, beam exposure time, geometry, etc.), the idea would certainly be rejected as not being in the best interests of the patient. Instead, a patient-specific treatment plan would be implemented in which treatment times are varied to deliver the same radiation dose to all patients. Patient-specific calculations of doses delivered to tumors and normal tissues have been routine in external beam radiotherapy and brachytherapy for decades. The routine use of a fixed GBq/kg, GBq/m$^2$, or simply GBq, administration of radionuclides for therapy is equivalent to treating all patients in external beam radiotherapy with the same protocol. Varying the treatment time to result in equal absorbed dose for external beam radiotherapy is equivalent to

accounting for the known variation in patients' uptake and retention half-time of activity of radio-nuclides to achieve equal tumor absorbed dose for internal-emitter radiotherapy. It has been suggested that fixed activity-administration protocol designs provide little useful information about the variability among patients relative to the normal organ dose than can be tolerated without dose-limiting toxicity compared to radiation dose-driven protocols.

In 2008, Stabin addressed the arguments commonly raised against performing patient-individualized dosimetry for nuclear medicine patients and offered reasoned arguments to refute them. The arguments remain and are the same; however, more data are available to show the usefulness of individualized dosimetry. He concluded that:

Treating all nuclear medicine patients with a single, uniform method of activity administration amounts to consciously choosing that these patients be treated with a lower standard of care than patients who receive radiation externally for cancer treatments.

It has been demonstrated that high-quality individualized therapy treatment planning can be provided for each patient. There are wide variations in tissue and tumor uptake and clearance between patients. Many centers can provide high-quality dosimetry for these treatments in a reliable and reproducible manner. "It's how we've always done it" is simply not an excuse to continue denying patients receiving radiopharmaceutical therapy the same quality of care that we give every day to patients receiving external beam radiation therapy. If centers will not do this voluntarily, the appropriate government agencies should make it mandatory. If the government will not act, patients and patient advocacy groups should force the issue by litigation, if necessary. It is time to end this traditional approach of all patients being treated with fixed-dosing approaches that ignore their individual characteristics and deny them the high quality of therapy that they deserve. In the following text, we will address some of the typical arguments made against performing patient-individualized dosimetry.

■ *Performing such calculations is difficult and expensive, requiring too much effort*—It is true that 3–5 planar or single photon emission computed tomography with computed tomography (SPECT/CT) or positron emission tomography (PET/CT) scans are needed to provide a patient-individualized treatment plan for nuclear medicine therapy, whereas a single CT study suffices for external beam therapy. However, these patients are dying of cancer, and they deserve the best therapy we can provide. They have been through a battery of difficult and invasive procedures, possibly surgeries, chemotherapy, etc., and will certainly consent to lying on an imaging table a few times to evaluate their biodistribution and biokinetics. We are tailoring all other kinds of medical procedures to individual patients to optimize each subject's treatments; we should certainly do this for therapy with high doses of radiation. Flux (2006) showed that the expense of performing dose calculations is not particularly high. The most basic kind of dose calculation, as was performed for $^{131}$I Bexxar therapy (before its removal from the market) with the gathering of whole-body retention at multiple time points, with simple regression analysis and estimation of only *whole-body* dose, takes perhaps half a day and costs around €200 per patient. Performing organ-based dose calculations, with 3–5 planar radionuclide scans with outlining of organ regions of interest, regression analysis of the individual organ curves, and calculation of mean organ dose using standardized dosimetry codes such as OLINDA/EXM (Stabin, Sparks, and Crowe 2005), takes perhaps one day of a physicist's time at a cost of perhaps €1,600 per patient. State-of-the-art individualized dosimetry, with 3D dose characterization of normal organ and tumor dose, using 3–5 SPECT scans, image registration, Monte Carlo analysis, and characterization of dose distributions and dose–volume histograms, may take up to 3 days of intensive analysis, with an approximate cost of €5,500. This is clearly a considerable cost, but it is not unlike the cost currently routinely accepted for performing intensity-modulated radiation therapy (IMRT), which is also estimated to cost around €5,500 per patient! Besides the better prognosis for the patient, which is the principal goal of therapy, the elimination of repeated therapy follow-up visits and other treatments is also economically attractive.

■ *No standardized methods exist for performing individualized dose calculations, and methods vary significantly among different institutions*—Dose calculations have been standardized for a number of years in OLINDA/EXM (Stabin et al. 2005) personal computer codes, which implement the methods outlined by the RADAR group of the Medical Internal Radiation Dose (MIRD) Committee of the Society of Nuclear Medicine (Stabin and Siegel 2003). These codes use the well-established standard models for reference adults, children, and pregnant women (Cristy and Eckerman 1987; Stabin et al. 1995), and have been widely employed in the international nuclear medicine community. These dose calculations are useful and nearly universally accepted in establishing standard doses for diagnostic radiopharmaceuticals for individuals of these fixed age and mass characteristics, which is needed in the drug approval process, for university approval committees to use in evaluating research proposals, and for other similar applications. Standardized methods for data gathering and reduction have been clearly summarized by Siegel et al. (1999), and protocols for obtaining high-quality data at sufficient time points to characterize dosimetry are easily and routinely

defined for individual radiopharmaceutical products.
Dewaraja et al. (2012) outlined detailed methods for obtaining quantitative data for dosimetry in SPECT; PET images
are always quantitative, as physicians are interested in
standard uptake values (SUVs), whereas SPECT images are
generally used only to obtain an image and not absolute
values of uptake. In these two documents, clear methods
are provided to obtain tissue and tumor uptakes from ante-
rior/posterior planar imaging, or tomographic images. The
number and spacing of time points is clearly explained.
Thus, anyone can obtain reasonable data for internal dose
calculations with an appropriate amount of attention to
detail. Different centers may use different approaches, to be
sure. Intercomparison exercises between centers have been
performed, with most centers having good success rates
at obtaining activity values within 10–20% of the known
answers.

■ *Dose calculations made to date have had poor success in
predicting tissue response*—The use of these standardized
procedures in therapy applications has allowed development of reproducible results, but, in some cases, the
results have not correlated well with observed effects in
some patient populations. Marrow toxicity from therapy
with internal emitters is manifested by hematological
changes in circulating platelets, lymphocytes, granulocytes,
reticulocytes, and red blood cells. Attempts to correlate
hematological toxicity with marrow dose, when marrow
cells are specifically targeted, have not been particularly
successful in the past, in part due to uncertainties in the
actual absorbed dose, but also due to the difficulty in
assessing marrow functional status prior to therapy (Siegel
et al. 1989, 1990; DeNardo et al. 1997; Eary et al. 1997; Lim
et al. 1997; Breitz, Fisher D, and Wessels 1998; Behr et al.
1999; Juweid et al. 1999). Dose–response analyses showed
that whole-body absorbed dose and red marrow absorbed
dose are often the best predictors of hematological toxicity,

as measured by platelet toxicity grade, with red marrow dose being slightly better. The correlations of a number of marrow toxicity indices with marrow dose for $^{90}$Y-labeled Zevalin, calculated using the reference adult phantom provided in the MIRDOSE code, with over 150 subjects, were disappointing. This led to the approval of the compound with no requirement for performing patient-individualized dose calculations. Hence, standardized methods and models are available and widely accepted, but refinements of these models are needed to implement patient-specific dosimetry that can be used in clinically relevant applications. However, others have shown good correlations between dose and tissue response, for normal tissues and tumors. However, notable successes have been demonstrated as well, when certain patient-individualized corrections are applied. Shen et al., using a $^{90}$Y-labeled antibody in radioimmunotherapy, obtained an $r$ value of 0.85 for correlation of marrow dose with observed marrow toxicity, using patient-specific marrow mass estimated from CT images, and through estimation of the total marrow mass from the mass of the marrow in three lumbar vertebrae (Figure 8.1) (Shen et al. 2002). Siegel et al. (2003) obtained a correlation coefficient of 0.86 between platelet nadir and calculated marrow dose but with an ingenious modification based on the levels of a stimulatory cytokine (FLT3-L) measurable in peripheral blood that indicates the possible present status of a subject's marrow, in the use of an $^{131}$I anti-carcinoembryonic antigen (CEA) antibody (Figure 8.2).

While others have failed to find firm correlations between tumor dose and observed response (Sgouros et al. 2003), Pauwels et al. (2005) found a convincing relationship, in their study of 22 patients with $^{90}$Y Octreother (an anti-somatostatin peptide) (Figure 8.3), employing PET imaging with $^{86}$Y to characterize the *in vivo* kinetics of the agent. Koral et al. (2000) showed that

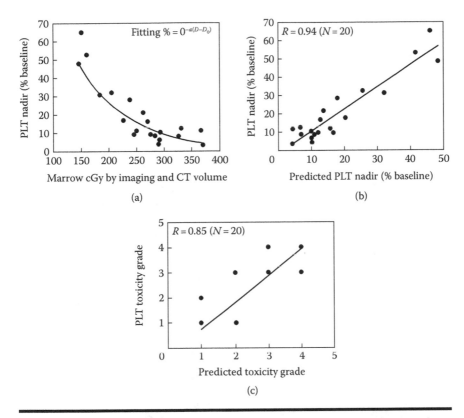

**Figure 8.1 Improved correlations between bone marrow dose and observed effects on marrow elements. (From Shen S et al., 2002, *J Nucl Med*, 43, 1245–1253.)**

tumor radiation dose was clearly linked with the likelihood of subjects to obtain a complete response to therapy. These patients were *de novo* to therapy, and thus the success of relating dose to response was higher than in many studies in the literature in which poor correlations between radiation dose and response have been found. Bone marrow reserve and tumor response are clearly complicated with previous radiation or chemotherapy; thus, the chances of finding strong relations between these variables are lower. Juweid et al. (1999) showed that bone marrow dose, calculated using standardized models with patient-specific correction of marrow mass, was the most important factor in predicting hematologic toxicity after radiotherapy of treatment of

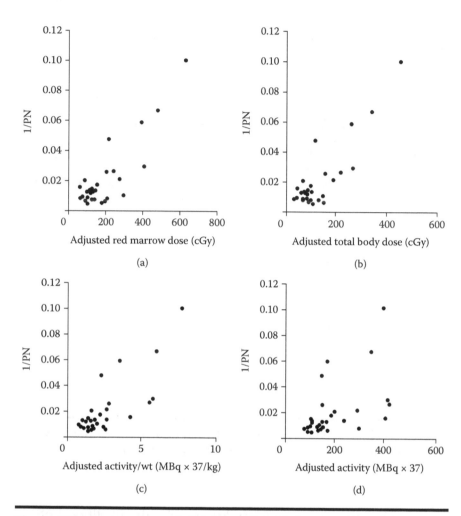

**Figure 8.2 Comparison of platelet nadir with FLT-3-L adjusted marrow dose: (a) Adjusted red marrow dose (cGy); (b) Adjusted total body dose (cGy); (c) Adjusted activity/wt (MBq x 37/kg); 9d) Adjusted activity (MBq x 37). (From Siegel JA et al., 2003, *J Nucl Med*, 44, 67–76.)**

carcinogenic embryonic antigen producing cancers who received [131]I labeled antibodies. Barone et al. (2005) noted the importance of the use of patient-individualized dose and dose-rate calculations, with patient-individualized kidney mass measurements, to predict kidney toxicity in the use of [90]Y (DOTATOC), again using [86]Y for quantitative imaging (Figure 8.4). Furthermore, recent evidence offered by some researchers indicates that the more

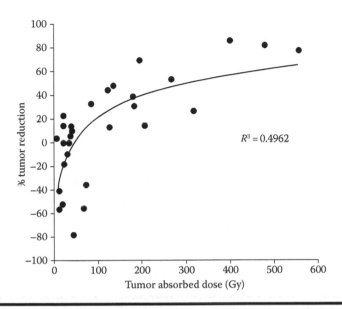

**Figure 8.3  Tumor dose response with ⁹⁰Y-DOTATOC. (From Pauwels S et al., 2005, *J Nucl Med*, 46, 92S–98S.)**

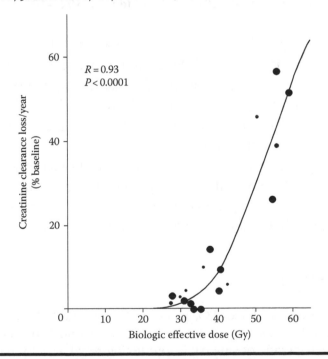

**Figure 8.4  Creatinine clearance as a function of renal absorbed doses. (From Barone R et al., 2005, *J Nucl Med*, 46, 99S–106S.)**

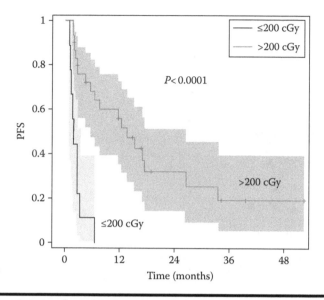

**Figure 8.5 Progression-free survival in lymphoma patients—left, dose <2 Gy, PFS 1.9 months; right, dose >2 Gy, PFS 13.6 months. (From Dewaraja YK et al., 2014, *J Nucl Med*, 55, 1047–1053.)**

sophisticated quantities, such as the biologically effective dose (BED), not just the absorbed dose, are the parameters that should be characterized on a patient-specific basis to explain observed biological effects, in both internal and external dose calculations (Bodey, Flux, and Evans 2003; Barone et al. 2005; Dale and Carabe-Fernandez 2005). Dewaraja et al. (2014) evaluated relationships between tumor dose and patient outcomes in radioimmunotherapy with [131]I for lymphoma, and found a clear difference in progression-free survival for patients receiving more than 2 Gy to their marrow (Figure 8.5).

# Bibliography

Barone R, Borson-Chazot FO, Valkerna R, Walrand S, Chauvin F, Gogou L, Kvols LK, Krenning EP, Jamar F, and Pauwels S. 2005. Patient-specific dosimetry in predicting renal toxicity with Y-90-DOTATOC: relevance of kidney volume and dose rate in finding a dose–effect relationship. *J Nucl Med*, 46, 99S–106S.

Behr TM, Sharkey RM, Juweid ME, Dunn RM, Vgg RC, Siegel JA, and Goldenberg DM. 1999. Hematological toxicity in the radio-immunotherapy of solid cancers with [131]I-labeled anti-CEA NP-4 IgG1: dependence on red marrow dosimetry and pretreatment. In *Proceedings of the Sixth International Radiopharmaceutical Dosimetry Symposium*, May 7–10, 1996, Vol I: 113–125. ORAU, Gatlinburg, TN Symposium.

Bodey RK, Flux GD, and Evans PM. 2003. Combining dosimetry for targeted radionuclide and external beam therapies using the biologically effective dose. *Cancer Biother Radiopharm*, 18(1), 89–97.

Breitz H, Fisher D, and Wessels B. 1998. Marrow toxicity and radiation absorbed dose estimates from rhenium-186-labeled monocolonal antibody. *J Nucl Med*, 39(10), 1746–1751.

Calvert AH, Newell DR, Gumbrell LA, O'Reilly S, Burnell M, Boxall FE, Siddik ZH, Judson IR, Gore ME, and Wiltshaw E. 1989. Carboplatin dosage: prospective evaluation of a simple formula based on renal function. *J Clin Oncol*, 7(11), 1748–1756.

Cristy M and Eckerman K. 1987. *Specific Absorbed Fractions of Energy at Various Ages from Internal Photons Sources*. ORNL/TM-8381 V1-V7. Oak Ridge National Laboratory, Oak Ridge, TN.

Dale R and Carabe-Fernandez A. 2005. The radiobiology of conventional radiotherapy and its application to radionuclide therapy. *Cancer Biother Radiopharm*, 20(1), 47–51.

DeNardo DA, DeNardo GL, O'Donnell RT, Lim S-M, Shen S, Yuan A, and DeNardo SJ. 1997. Imaging for improved prediction of myelotoxicity after radioimmunotherapy. *Cancer*, 80, 2558–2566.

Dewaraja YK, Frey EC, Sgouros G, Brill AB, Roberson P, Zanzonico PB, and Ljungberg M. 2012. MIRD pamphlet no. 23: quantitative SPECT for patient-specific 3-dimensional dosimetry in internal radionuclide therapy. *J Nucl Med*, 53(8), 1310–1325.

Dewaraja YK, Schipper MJ, Shen J, Smith LB, Murgic J, Savas H, Youssef E, et al. 2014. Tumor-absorbed dose predicts progression-free survival following [131]I-Tositumomab radioimmunotherapy. *J Nucl Med*, 55, 1047–1053.

Eary JF, Krohn KA, Press OWQ, Durack L, and Bernstein ID. 1997. Importance of pre-treatment radiation absorbed dose estimation for radioimmunotherapy of non-Hodgkin's lymphoma. *Nucl Med Biol*, 24, 635–638.

Flux G. 2006. *Cost Estimates in this Section from Personal Communication*, Royal Marsden Hospital, UK.

Hoskins PJ, Swenerton KD, Pike JA, Wong F, Lim P, Acquino-Parsons C, and Lee N. 2001. Paclitaxel and Carboplatin, Alone or With Irradiation, in Advanced or Recurrent Endometrial Cancer: A Phase II Study. *J Clin Oncol*, 19(20), 4048–4053.

Juweid ME, Zhang C-H, Blumenthal RD, Hajjar G, Sharkey RM, and Goldenberg DM. 1999. Prediction of hematologic toxicity after radioimmunotherapy with [131]I labeled anti-carcinoembryonic antigen monoclonal antibodies. *J Nucl Med*, 40(10), 1609–1616.

Koral KF, Dewaraja Y, Clarke LA, Li J, Zasadny R, Rommelfanger SG, Francis IR, Kaminski MS, and Wahl RL. 2000. Tumor absorbed dose estimates versus response in tositumomab therapy of previously untreated patients with follicular non-Hodgkin's lymphoma: preliminary report. *Cancer Biother Radiopharm*, 15(4), 347–355.

Lim S-M, DeNardo GL, DeNardo DA, Shen S, Yuan A, O'Donnell RT, and DeNardo SJ. 1997. Prediction of myelotoxicity using radiation doses to marrow from body, blood and marrow sources. *J Nucl Med*, 38, 1374–1378.

Morgan G, Wardy R, and Bartonz M. 2004. The Contribution of Cytotoxic Chemotherapy to 5-year Survival in Adult Malignancies. *J Clin Oncol*, 16, 549–560.

Newell DR, Pearson ADJ, Balmanno K, Price L, Wyllie RA, Keir M, Calvert AH, Lewis IJ, Pinkerton CR, and Stevens MC. 1993. Carboplatin pharmacokinetics in children: the development of a pediatric dosing formula. *J Clin Oncol*, 11, 2314–2323.

Pauwels S, Barone R, Walrand S, Borson-Chazot F, Valkema R, Kvols LK, Krenning EP, and Jamar F. 2005. Practical dosimetry of peptide receptor radionuclide therapy with [90]Y-labeled somatostatin analogs. *J Nucl Med*, 46(1), 92S–98S.

Sgouros G, Squeri S, Ballangrud AM, Kolbert KS, Teitcher JB, Panageas KS, Finn RD, Divgi CR, Larson SM, and Zelenetz AD. 2003. Patient-specific, 3-dimensional dosimetry in non-Hodgkin's lymphoma patients treated with [131]I-anti-B1 antibody: assessment of tumor dose-response. *J Nucl Med*, 44, 260–268.

Shen S, Meredith RF, Duan J, Macey DJ, Khazaeli MB, Robert F, and LoBuglio AF. 2002. Improved prediction of myelotoxicity using a patient-specific imaging dose estimate for non-marrow-targeting [90]Y-antibody therapy. *J Nucl Med*, 43, 1245–1253.

Siegel JA, Lee RE, Pawlyk DA, Horowitz JA, Sharkey RM, and Goldenberg DM. 1989. Sacral scintigraphy for bone marrow dosimetry in radioimmunotherapy. *Int J Rad Appl Instrum (B)*, 16, 553–559.

Siegel JA, Stabin MG, and Brill AB. 2002. The importance of patient-specific radiation dose calculations for the administration of radionuclides in therapy. *Cell Mol Biol* (Noisy-le-grand), 48(5), 451–459.

Siegel JA, Thomas S, Stubbs J, Stabin M, Hays M, Koral K, Robertson J, Howell R, Wessels B, Fisher D, Weber D, and Brill A. 1999. MIRD pamphlet no 16: techniques for quantitative radiopharmaceutical biodistribution data acquisition and analysis for use in human radiation dose estimates. *J Nucl Med*, 40, 37S–61S.

Siegel JA, Wessels BW, Watson EE, Stabin MG, Vriesendorp HM, Bradley EW, Badger CC, Brill AB, and Kwok CS. 1990. Bone marrow dosimetry and toxicity for radioimmunotherapy. *Antibody, Immunoconjugates, and Radiopharmaceuticals*, 3, 213–233.

Siegel JA, Yeldell D, Goldenberg DM, Stabin MG, Sparks RB, Sharkey RM, Brenner A, and Blumenthal RD. 2003. Red marrow radiation dose adjustment using plasma flt3-l cytokine levels: improved correlations between hematologic toxicity and bone marrow dose for radioimmunotherapy patients. *J Nucl Med*, 44, 67–76.

Stabin MG. 2008. The case for patient-specific dosimetry in radionuclide therapy. *Cancer Biother Radio*, 23(3), 273–284.

Stabin MG and Siegel JA. 2003. Physical Models and dose factors for use in internal dose assessment. *Health Phys*, 85(3), 294–310.

Stabin MG, Sparks RB, and Crowe E. 2005. OLINDA/EXM: The second-generation personal computer software for internal dose assessment in nuclear medicine. *J Nucl Med*, 46, 1023–1027.

Stabin MG, Xu XG, Emmons MA, Segars WP, Shi C, and Fernald MJ. 2012. RADAR reference adult, pediatric, and pregnant female phantom series for internal and external dosimetry. *J Nucl Med*, 53, 1807–1813.

Stabin M, Watson E, Cristy M, Ryman J, Eckerman K, Davis J, Marshall D, and Gehlen K. 1995. *Mathematical Models and Specific Absorbed Fractions of Photon Energy in the Nonpregnant Adult Female and at the End of Each Trimester of Pregnancy*. ORNL Report ORNL/TM-12907.

Wahl R. 2005. Tositumomab and 131I Therapy in Non-Hodgkin's Lymphoma. *J Nucl Med*, 46(1), 128S–140S.

# Chapter 9

# Radiation Biology

It is well known that ionizing radiation can cause damage to biological systems. The discovery of radiation and its use in medical applications caused great excitement, and many investigations began in the late 1800s and early 1900s. Very soon, however, deleterious effects were observed, including damage to the skin, induction of cancer, and the particularly tragic episode of the radium dial painters, who ingested significant quantities of $^{226}$Ra in factories in which luminous dial watches were made (Mullner 1989), as they had the habit of sharpening the tips of their paint brushes with their lips. They suffered several types of bone cancer, including carcinomas of the paranasal sinuses, which are particularly rare and clearly due to their exposure to radium. They even manifested spontaneous fractures in their spines and jaws; others died of anemia. They were compensated eventually, although the factory owners denied responsibility for as long as they could.

Our classical ideas about how radiation causes damage to biological organisms is being challenged currently by some new and fascinating observations, which will be discussed in the following text. All of the mechanisms for these effects are not currently understood, so we will look at the findings, but full explanation of these effects awaits further research and investigation of the pathways for the effects. First, we will discuss the classical view of radiation damage to cells and tissues, which is certainly not incorrect, but perhaps incomplete.

# Types of Radiation-Induced Effects

Two broad categories of radiation-related effects in humans are identified as *nonstochastic* and *stochastic*. The former occurs soon after exposure to radiation, and the latter appears at times long after the exposure, at least 5 years, and typically much longer than that. They have fundamental characteristics that distinguish them, as discussed in the following text.

## Nonstochastic Effects

*Nonstochastic effects* (now officially called *deterministic effects*, previously also called *acute effects*) are generally observed soon after exposure to radiation. As they are *nonstochastic* (nonprobabilistic) in nature, they will always be observed (if a certain dose threshold is exceeded), and there is generally no doubt that they were caused by radiation exposure. The major identifying characteristics of nonstochastic effects are as follows:

1. There is a threshold dose below which the effects will not be observed.
2. Above this threshold, the magnitude of the effect increases with dose.
3. The effect is clearly associated with radiation exposure.

Examples:

■ Erythema (reddening of the skin)
■ Epilation (loss of hair)

- Depression of bone marrow cell division (observed in counts of formed elements in peripheral blood)
- Nausea, vomiting, diarrhea (the *NVD* syndrome)—often observed in victims after acute exposure to radiation
- Central nervous system damage
- Damage to the unborn child—physical deformities, micro-cephaly (small head size at birth), mental retardation

When discussing nonstochastic effects, it is important to note that some organs are more radiosensitive than others. The so-called *Law of Bergonie and Tribondeau* (Bergonie and Tribondeau 1906) states that cells tend to be radiosensitive if they have three properties:

- Cells have a high division rate.
- Cells have a long dividing future.
- Cells are of an unspecialized type.

One way of stating the law might be to say: "The radio-sensitivity of a cell type is proportional to its rate of division and inversely proportional to its degree of specialization." Hence, rapidly dividing and unspecialized cells, as a rule, are the most radiosensitive. Two important examples are pro-genitor cells in the bone marrow and most cells in the developing embryo/fetus. In the case of marrow, the progenitor cells, through many generations of cell division, produce a variety of different functional cells that are very specialized (e.g., red blood cells, lymphocytes, leukocytes, and platelets) (Figure 9.1). Some of these functional cells do not divide at all and are thus themselves quite radioresistant. However, if the marrow receives a high dose of radiation, damage to these progenitor cells is very important to the health of the organism. As we will see shortly, if these cells are affected, this will be manifested in a short period in a measurable decrease in the number of formed elements in the peripheral blood. If the damage is severe enough, the person may not survive. If not,

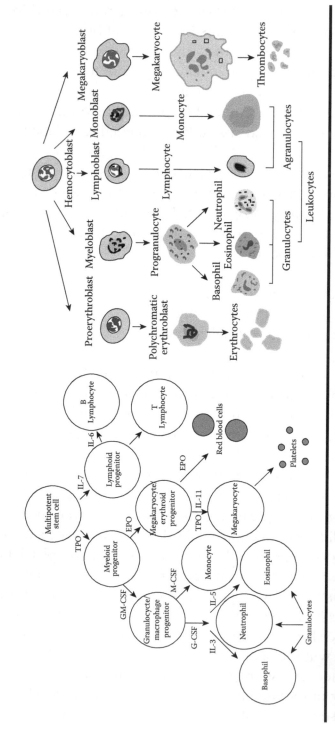

**Figure 9.1** Generations of red marrow progenitor cells. Available from http://stemcell7.blogspot.com/2007/06/stem-cell-2.html

the progenitor cells will eventually repopulate and begin to replenish the numbers of the formed elements, and subsequent blood samples will show this recovery process.

In the fetus, organs and systems develop at different rates. At the moment of conception, of course, we have one completely undifferentiated cell that becomes two cells after one division, then four, then eight, and so on. As the rapid cell division proceeds, groups of cells "receive their assignments" and differentiate to form organs and organ systems, still with a very rapid rate of cell division. At some point, individual organs become well defined and formed, and cell division slows as the fetus simply adds mass. However, while differentiation and early rapid cell division is occurring, these cells are quite radiosensitive, and a high dose to the fetus may cause fetal death, or damage to individual fetal structures. This is discussed further in the following text. On the other hand, in an adult, cells of the central nervous system (CNS) (brain tissue, spinal cord, etc.) are very highly specialized and have very low, or no, rate of division. The CNS is thus particularly radioresistant. One important nonstochastic effect is death. This results from damage to the bone marrow (first), then to the gastrointestinal tract, and then to the nervous system.

## Stochastic Effects

*Stochastic effects* are effects that are, as the name implies, probabilistic. They may or may not occur in any given exposed individual. These effects generally manifest for many years, even decades after the radiation exposure (and were once called *late effects*). The major characteristics of stochastic effects, in direct contrast with those for nonstochastic effects, are as follows:

1. A threshold may not be observed.
2. The probability of the effect increases with dose.
3. You cannot definitively associate the effect with radiation exposure.

Examples include:

- *Cancer induction*—The most important and widely discussed stochastic effect is cancer. It is well known that ionizing radiation causes cancer at high doses and dose rates. A considerable controversy remains about whether this is also true at low doses and dose rates, as will be discussed in the following text. A number of human and animal populations have been studied to try and develop causative links and quantitative relationships describing the effects of radiation exposure for different cancer types. The first complication is that a radiation-induced cancer cannot be distinguished from cancers that occur in the exposed population in the absence of any radiation exposure; thus, *excess* cancers in a population are sought, and the question of the statistical significance of a number of excess cancers (or even levels of cancer below the *background* cancer rate) is always present. The most important population studied to date to study relationships between radiation exposures and cancer induction (or promotion) is the population of survivors of the atomic bomb attacks in Hiroshima and Nagasaki, Japan, at the end of World War II. Several hundred thousand people died either instantly or within the first year after the attacks, from physical injuries and radiation sickness. The surviving population has been extensively studied over the years following the attacks. The main institution in this follow-up effort is the Radiation Effects Research Foundation (RERF), with locations in both Hiroshima and Nagasaki. The RERF (formerly the Atomic Bomb Casualty Commission) was founded in April 1975 and is a private nonprofit Japanese foundation. Some 36,500 survivors who were exposed beyond 2.5 km have been medically followed continuously since the blasts. Of this population, about 4,900 cancer deaths have been identified, including about 180 leukemia deaths and 4,700 deaths from cancers other than leukemia—of these, only about 89 and 340 deaths, respectively, appear to be attributable to radiation. There are also published data for a number of populations of individuals exposed to various medical studies using high levels of radiation (principally from the early 1900s).

Important types of cancer that have been well studied include (Stabin 2007):

– *Leukemia*: Acute myelogenous leukemia (AML) and, to a lesser extent, chronic myelogenous or acute lymphocytic leukemia are among the most likely forms of malignancy resulting from whole-body exposure to radiation. Leukemia has the shortest latent period (time between radiation exposure and expression of cancer) of all forms of cancer. Increases in leukemia incidence have been clearly shown in a number of populations exposed to high levels of radiation, including:
  • Early radiologists
  • Japanese bombing survivors
  • British children irradiated *in utero*
– *Bone cancer*: The Radium Dial painters expressed a number of different osseous malignancies, in addition to their acute damages to their skeletons. Interestingly, no significant incidence of bone cancer has been observed in the Japanese bomb survivors.
– *Lung cancer*: It has been clearly seen in the Japanese bomb survivors, but was first observed historically, as early as the sixteenth century, in populations of uranium miners exposed to high levels of the radioactive gas $^{222}$Rn and its progeny. This species is a member of the $^{238}$U decay series. As with leukemia and many solid tumors, it is clear that exposure to high levels of $^{222}$Rn and its progeny causes lung cancer, and some have raised concerns about much lower levels

of $^{222}$Rn in residences. The Environmental Protection Agency of the United States clearly advocates that exposures to radon follow an LNT model and suggest remediation of people's homes to reduce radon levels above certain levels (Figure 9.2).

– *Thyroid cancer*: It has been clearly associated with radiation exposure. Children irradiated with x-rays for enlarged thymus glands, ringworm of the scalp, and acne (again in the early days of radiation use) showed significant rates of thyroid cancer. In 1954, an aboveground test of a nuclear weapon in the Pacific Ocean went awry, with weather conditions not matching expectations, and the weapon yield was considerably greater than expected. A number of people living on the nearby Rongerik, Rongelap, Ailinginae, and Utirik atolls (the "Marshallese Islanders") were blanketed in fallout from the weapon, resulting in considerable whole-body exposures, skin lesions from radiation exposure, and intakes of $^{131}$I. In April 1986, the most significant disaster in the nuclear power industry's history occurred at the Chernobyl nuclear power plant in the Ukraine. The event was the product of a flawed Soviet reactor design coupled with serious mistakes made by the poorly trained plant operators. The accident destroyed one reactor and killed 31 people, 28 due to their radiation exposures. Another 100–150 cases of acute radiation poisoning were confirmed (the subjects recovered). There were no off-site acute radiation effects, but large areas of Belarus, Ukraine, Russia, as well as areas of Eastern Europe were contaminated to varying degrees. Many cases of thyroid cancer in children and adolescents in the contaminated areas were linked to their exposure to the $^{131}$I releases from the site (other shorter-lived isotopes of iodine may also be involved in the radiation doses received by the children's thyroids).

**Radon Risk If You've Never Smoked**

| Radon level | If 1,000 people who never smoked were exposed to this level over a lifetime*... | The risk of cancer from radon exposure compares to**... | WHAT TO DO: |
|---|---|---|---|
| 20 pCi/L | About 36 people could get lung cancer | 35 times the risk of drowning | Fix your home |
| 10 pCi/L | About 18 people could get lung cancer | 20 times the risk of dying in a home fire | Fix your home |
| 8 pCi/L | About 15 people could get lung cancer | 4 times the risk of dying in a fall | Fix your home |
| 4 pCi/L | About 7 people could get lung cancer | The risk of dying in a car crash | Fix your home |
| 2 pCi/L | About 4 person could get lung cancer | The risk of dying from poison | Consider fixing between 2 and 4 pCi/L |
| 1.3 pCi/L | About 2 people could get lung cancer | (Average indoor radon level) | (Reducing radon levels below 2 pCi/L is difficult.) |
| 0.4 pCi/L | | (Average outdoor radon level) | |

**Radon Risk If You Smoke**

| Radon level | If 1,000 people who smoked were exposed to this level over a lifetime*... | The risk of cancer from radon exposure compares to**... | WHAT TO DO: Stop smoking and... |
|---|---|---|---|
| 20 pCi/L | About 260 people could get lung cancer | 250 times the risk of drowning | Fix your home |
| 10 pCi/L | About 150 people could get lung cancer | 200 times the risk of dying in a home fire | Fix your home |
| 8 pCi/L | About 120 people could get lung cancer | 30 times the risk of dying in a fall | Fix your home |
| 4 pCi/L | About 62 people could get lung cancer | 5 times the risk of dying in a car crash | Fix your home |
| 2 pCi/L | About 32 people could get lung cancer | 6 times the risk of dying from poison | Consider fixing between 2 and 4 pCi/L |
| 1.3 pCi/L | About 20 people could get lung cancer | (Average indoor radon level) | (Reducing radon levels below 2 pCi/L is difficult.) |
| 0.4 pCi/L | About 3 people could get lung cancer | (Average outdoor radon level) | |

**Figure 9.2   EPA recommendations for remediation of homes against radon. (From EPA, 2015, http://www2.epa.gov/radon/health-risk-radon.)**

For different types of cancer, for doses above 100–200 mGy, a plot of the number of cancers observed against the dose received shows a clear upward trend. The question that none of the data sets clearly answer is that of the shape of the curve at low doses and dose rates. Many influential scientific advisory bodies, and many regulatory agencies, favor the so-called linear, no-threshold (LNT) model of radiation carcinogenesis. This model assumes that *any* radiation exposure, no matter how small, carries some risk of cancer induction. Others cite data that appear to demonstrate a threshold dose for cancer induction. There is even evidence to support the controversial theory of *hormesis*—that exposure to low levels of radiation is associated with less cancer induction than in systems deprived of all radiation exposure (Figure 9.3). The possible mechanism here is that exposure to radiation stimulates cellular repair mechanisms (the mechanism of *adaptive response*). In different experiments, hormetic and adaptive response mechanisms have been demonstrated.

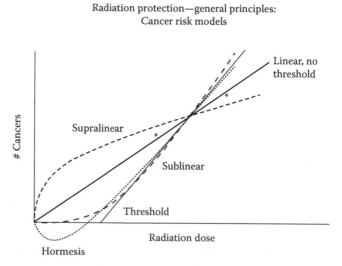

Radiation protection—general principles:
Cancer risk models

The two "*" symbols do not represent real data, just an indication that data at high enough doses show a positive link to radiation exposure with an upward trend with dose.

**Figure 9.3   Possible models of cancer induction as a function of radiation dose received.**

Doss (2013) recently performed an extensive review of much data related to the LNT model and hormesis and concluded that:

> The conclusion of zero threshold dose for carcinogenic effects of radiation in the recent updated report on the atomic bomb survivor cancer mortality data appears to be unjustified and may be the result of the restrictive functional forms that were used to fit the data. Also, the shape of the dose–response observed in the recent update of atomic bomb survivor data is clearly non-linear with the significant reduction in cancer mortality rate in the dose range of 0.3 Gy to 0.7 Gy.

The Biological Effects of Ionizing Radiation (BEIR) Committee of the National Academy of Sciences (NAS) concluded in its 2006 report (BEIR 2006) that:

> The scientific research base shows that there is no threshold of exposure below which low levels of ionizing radiation can be demonstrated to be harmless or beneficial.

At about the same time, and studying the same data, the French Academy of Sciences (FAS 2005) concluded that:

> In conclusion, this report raises doubts on the validity of using LNT for evaluating the carcinogenic risk of low doses (<100 mSv) and even more for very low doses (<10 mSv). The LNT concept can be a useful pragmatic tool for assessing rules in radioprotection for doses above 10 mSv; however, since it is not based on biological concepts of our current knowledge, it should not be used without precaution for assessing by extrapolation the risks associated with low and, even more so, with very low doses.

Siegel and Welsh (2016) reviewed the low-dose data from the Japanese bomb survivors and concluded that:

> Low-dose radiation exposure (<100–200 mSv) is likely beneficial, not harmful. The LNT model does not account for the body's response (i.e., repair) of any radiation-induced damage and assumes that all outcomes are linear across the range of high- and low-dose exposures; an assumption that is likely false. The use of the LNT model is not conservative and therefore should be abandoned …

They also noted the significant loss of life due to forced evacuations following the Fukushima accident (more than 1,000 deaths) due to radiophobia related to the LNT hypothesis.

- *Genetic effects (offspring of irradiated individuals)*— Radiation-induced hereditary effects have been clearly demonstrable in animal experiments involving mice and fruit flies, but *never* in any human population with statistical significance, including the Japanese bomb survivors, medical populations, and populations affected by the Chernobyl disaster. As with cancer, there is a spontaneous rate of mutations that is ongoing in the human population with no excess exposure to chemicals, radiation, or other mutagenic agents.

# Mechanisms of Radiation Damage

Radiation action on cells is described as *direct* or *indirect*. The former refers to when the DNA molecule has an interaction with an ionizing particle and suffers damage as a result. The latter, which are known to be more important overall (Kassis, Harapanhalli, and Adelstein 1999), involve the creation of ionized and excited molecules via interaction with water. These species are very interactive, seeking to return to their normal state, and can interact with the DNA molecule, causing damage. The radiation interaction occurs in four defined stages:

1. Physical
2. Prechemical
3. Early chemical
4. Late chemical

In the *physical* stage of water radiolysis (duration $10^{-15}$ s), a primary charged particle interacts through elastic and inelastic collisions. Inelastic collisions result in the ionization and excitation of water molecules, leaving behind ionized ($H_2O^+$) and excited ($H_2O^*$) molecules, and unbound subexcitation electrons ($e_{-sub}$). A subexcitation electron is one whose energy is not high enough to produce further electronic transitions. By contrast, some electrons produced in the interaction of the primary charged particle with the water molecules may have sufficient energy themselves to produce additional electronic transitions. These electrons may produce secondary track structures (called *delta rays*), beyond that produced by the primary particle. All charged particles can interact with electrons in the water both individually and collectively in the condensed, liquid, phase. The prechemical phase (from $10^{-15}$ s until about $10^{-12}$ s) involves some initial reactions and rearrangements of these species. If a water molecule is ionized, this results in the creation of an ionized water molecule and a free electron. The free electron rapidly attracts other water molecules, as the slightly polar molecule has a positive and negative pole, and the positive pole is attracted to the electron. A group of water molecules thus clusters around the electron, and it is known as a *hydrated electron* and is designated as $e_{aq}-$. The water molecule dissociates immediately:

$$H_2O \rightarrow H_2O^+ + e_{aq}- \rightarrow H + +OH^• + e_{aq}-$$

In an excitation event, an electron in the molecule is raised to a higher energy level. This electron may simply return to its original state, or the molecule may break up into an H and an OH radical (a radical is a species that has an unpaired electron in one of its orbitals; the species is not necessarily charged, but is highly reactive).

$$H_2O^* \rightarrow H^{\bullet} + OH^{\bullet}$$

The free radical species and the hydrated electron undergo many reactions with each other and other molecules in the system. Reactions with other commonly encountered molecules in aqueous systems are shown in the following table. These reactions and those with other molecular species have been studied and modeled by various investigators as well (Wright et al. 1985; Turner et al. 1994; Becker et al. 1997; Pimblott and LaVerne 1997). The reactions and their products have defined rate constants and *reaction radii* (the distance between the centers of two particles at which a reaction is assumed to occur):

| Reaction | $k\ (10^{10})\ M^{-1}\ s^{-1}$ | $R\ (nm)$ |
|---|---|---|
| $H^{\bullet} + OH^{\bullet} \rightarrow H_2O$ | 2.0 | 0.43 |
| $e_{aq}^- + OH^{\bullet} \rightarrow OH^-$ | 3.0 | 0.72 |
| $e_{aq}^- + H^{\bullet} + H_2O \rightarrow H_2 + OH^-$ | 2.5 | 0.45 |
| $e_{aq}^- + H_3O^+ \rightarrow H^{\bullet} + H_2O$ | 2.2 | 0.39 |
| $H^{\bullet} + H^{\bullet} \rightarrow H_2$ | 1.0 | 0.23 |
| $OH^{\bullet} + OH^{\bullet} \rightarrow H_2O_2$ | 0.55 | 0.55 |
| $2e_{aq}^- + 2H_2O \rightarrow H_2 + 2OH^-$ | 0.5 | 0.18 |
| $H_3O^+ + OH^- \rightarrow 2H_2O$ | 14.3 | 1.58 |
| $e_{aq}^- + H_2O_2 \rightarrow OH^- + OH^{\bullet}$ | 1.2 | 0.57 |
| $OH^{\bullet} + OH^- \rightarrow H_2O + O^-$ | 1.2 | 0.36 |

In the *early chemical* phase, the formed species diffuse and react with each other and possibly other molecules in the solution ($10^{-12}$ s until about $10^{-6}$ s). After this point, the track structure of the ionizing particle is mostly lost (Figure 9.2), and further interactions are unlikely. After $10^{-6}$ s, the *late chemical* stage, modeling of further interactions can be developed using differential rate-equation systems.

Damage to the cell may come in the form of single DNA strand breaks, which are more easily repaired through relation to the base pair complement on the opposite strand. Double DNA strand breaks are more difficult to repair, but cells have sophisticated repair mechanisms that can undo this damage. However, double strand breaks are more likely to result in cell killing, carcinogenesis, or mutation (Hall 2011).

*Cell survival studies* carried out *in vitro* allow study of cell survival of different kinds of cell lines. Several cell lines are available to researchers all over the world, to facilitate reproducibility of results. Radiation with high linear energy transfer (e.g., alpha particles) will exhibit a survival curve that is well characterized by a single exponential term.

## Challenges to the Classic Model

As noted at the beginning of the chapter, the classic model of DNA damage due to the indirect action of ionizing radiation, with complete or incomplete repair determining the fate of the cell, has been challenged recently by some new and surprising experimental data. The survival ($S$) at dose $D$, relative to the original number of cells ($S_0$), is given as $S = S_0\, e^{-D/D_0}$, where $D_0$ is the negative reciprocal of the slope of the straight-line curve (plotted on a semilog scale), and is called the *mean lethal dose*. When cells receive dose $D_0$, the surviving fraction is 0.37, which is $1/e$. Irradiation with low-low Linear Energy Transfer (LET) radiations will typically result in a curve that becomes linear at higher doses, but with a curvilinear behavior or *shoulder* at lower doses: $S = S_0 \left[ 1 - \left( 1 - e^{D/D_0} \right)^n \right]$. This is the *multitarget model*, in which the factor $n$ is related to the width of the *shoulder*. Another model is the *linear quadratic* model, given as $S = e^{-\alpha D}$ for high-LET radiations and $S = e^{-\alpha D - \beta d^2}$ for low-LET radiations. However, it is clear that, as much as these models predict many of the effects we see from ionizing radiation, there is new evidence that more complex mechanisms are ongoing in *tissues* and that cell-to-cell communication by some means can cause cell death in unirradiated cells, or stimulation of repair.

■ Hall (2003) described experimental data for *bystander effects* in two kinds of experiments—medium transfer and microbeam irradiation. In the former, cells are irradiated, then removed from the supporting medium, and unirradiated cells are placed into this medium and demonstrate a biological effect. In the latter, cells are plated, and microbeams of radiation irradiate only certain cells and not others, but the unirradiated cells manifest radiation effects.

■ Brooks (2004) cited remarkable evidence in the irradiation of partial lung volumes in rats. Irradiation of the lung base in rats resulted in a marked increase in the frequency of micronuclei in the shielded lung apex. However, radiation of the lung apex did not result in an increase in the chromosome damage in the shielded lung base. This suggests that a factor was transferred from the exposed portion of the lung to the shielded part and that this transfer has direction from the base to the apex of the lung. In another experiment, exposure of the left lung resulted in a marked increase in micronuclei in the unexposed right lung.

■ Hall (2003) showed a bystander effect in V79 cells irradiated *in vitro* with microbeams of alpha particles. Cells that were irradiated naturally showed decreased survival dependent on the number of alpha particles to which they were exposed, but cells that were not hit with *any* alpha particles also showed decreased survival (Figure 9.4).

■ More remarkably, Sawant et al. (2001) showed similar bystander effects in C3H 10T1/2 cells in culture, but the bystander effect was diminished if cells were exposed to 2 cGy of gamma radiation 6 h before being irradiated with alpha particle microbeams! (Figure 9.5).

■ Mairs et al. (2007) demonstrated bystander effects in medium transfer experiments using high- and low-LET radiations to irradiate cells and then exposed unirradiated cells to the media from the irradiated cells. The results were surprising—a clear bystander effect was seen, which reached a plateau after a certain dose for gamma rays, but not for $^{131}I$ beta particles. Even more surprising, the effect for $^{123}I$ Auger electrons and $^{211}At$ alpha particles had a "U" shape, that is, the effect increased with dose to a point, but, at higher doses, the effect was diminished (Figure 9.6)!

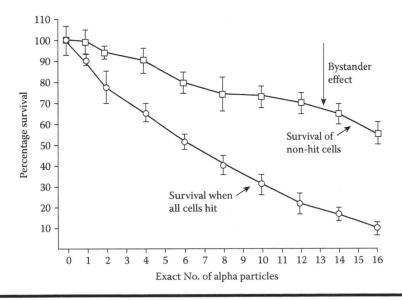

**Figure 9.4  Bystander effect shown by Hall (2003) in V79 cells in which all cells or 10% of cells were struck by 1–16 alpha particles.**

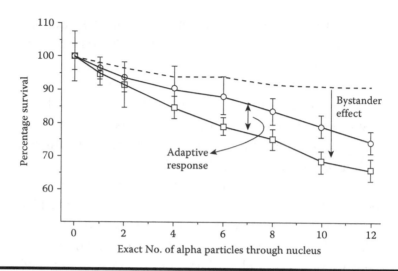

**Figure 9.5** **Bystander effect in microbeam irradiation experiments is diminished by irradiating the cells with low doses of gamma rays. (From Sawant SG et al., 2001, *Radiat Res*, 156, 177–180.)**

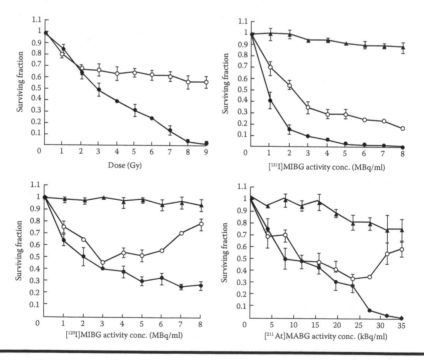

**Figure 9.6** **Bystander effect in medium transfer experiments with different kinds of radiation. (From Mairs RJ et al., 2007, *Dose-Response*, 5, 204–213.)**

■ In perhaps the most interesting study of all, Kishikawa et al. (2006) performed a medium transfer study using two different isotopes of iodine, $^{123}$I and $^{125}$I, in an *in vivo* experiment and observed bystander effects that either resulted in faster or slower tumor growth in nude mice. Cells were lethally irradiated, and unirradiated cells were placed in the supernatant and then injected into the mice. The cells exposed to the supernatant of the $^{125}$I-irradiated cells grew slower than controls, which was expected, but the cells exposed to the supernatant of the $^{123}$I-irradiated cells grew faster than controls (Figure 9.7)! They identified

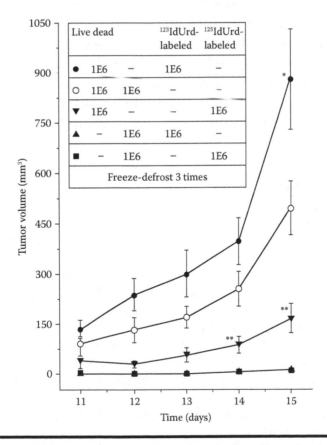

**Figure 9.7** *Inhibitory and Stimulatory* **bystander effects in medium transfer experiments** *in vivo.* **(From Kishikawa H et al., 2006,** *Radiat Res,* **165, 688–694.)**

different compounds in the supernatants of the two cell groups, but did not definitively explain the results mechanistically.

We are now realizing that:

■ Cells do not respond to radiation, tissues do.
■ Cells are signaling one another in reaction to radiation damage.

Our understanding of radiation biology is evolving; our understanding of radiation effects on DNA is not incorrect, but seems to be only part of the picture of how radiation affects biological systems. More research is needed into the mechanisms by which cells and tissues respond to interactions with ionizing radiation. Our understanding will necessarily mature and change, and radiation biology textbooks in a few years may be quite different from those we have today.

# Bibliography

Becker D, Sevilla MD, Wang W, and LaVere T. 1997. The role of waters of hydration in direct-effect radiation damage to DNA. *Radiat Res*, 148, 508–510.

BEIR (Committee to Assess Health Risks from Exposure to Low Levels of Ionizing Radiation). Beir VII Phase 2. 2006. *National Research Council of the National Academies*. The National Academies Press, Washington, DC.

Bergonie J and Tribondeau L. 1906. De quelques resultats de la Radiotherapie, et esaie de fixation d'une technique rationelle. *Comptes Rendu des Seances de l'Academie des Sciences*, 143, 983–985.

Brooks AL. 2004. Evidence for "bystander effects" in vivo. *Hum Exp Toxicol*, 23, 67–70.

Doss M. 2013. Linear no-threshold model vs. radiation hormesis. *Dose-Response*, 11, 495–512.

EPA (Environmental Protection Agency). 2015. http://www2.epa.gov/radon/health-risk-radon.

FAS (French Academy of Sciences). 2005. Dose–effect relationships and estimation of the carcinogenic effects of low doses of ionizing radiation. *Academie des Sciences (Academy of Sciences)—Academie nationale de Medecine (National Academy of Medicine)*.

Hall EJ. 2003. The bystander effect. *Health Phys*, 85(1), 31–35.

Hall EJ. 2011. *Radiobiology for the Radiologist*. Lippincott Williams Wilkins, Philadelphia PA.

Kassis AI, Harapanhalli RS, and Adelstein SJ. 1999. Strand breaks in plasmid DNA after positional changes of Auger electron-emitting iodine-125: direct compared to indirect effects. *Radiat Res*, 152(5), 530–538.

Kishikawa H, Wang K, Adelstein SJ, and Kassis AI. 2006. Inhibitory and stimulatory bystander effects are differentially induced by iodine-125 and iodine-123. *Radiat Res*, 165, 688–694.

Mairs RJ, Fullerton NE, Zalutsky MR, and Boyd M. 2007. Targeted radiotherapy: microgray doses and the bystander effect. *Dose-Response*, 5, 204–213.

Mullner R. 1989. *Deadly Glow: The Radium Dial Worker Tragedy*. American Public Health Association, Washington, DC.

Pimblott SM and LaVerne JA. 1997. Stochastic simulation of the electron radiolysis of water and aqueous solutions. *J Phys Chem A*, 101, 5828–5838.

Sawant SG, Randers-Pehrson G, Metting NF, and Hall EJ. 2001. Adaptive response and the bystander effect induced by radiation in C3H 10T1/2 cells in culture. *Radiat Res*, 156, 177–180.

Siegel JA and Welsh JS. 2016. Does imaging technology cause cancer? Debunking the linear no-threshold model of radiation carcinogenesis. *Technol Cancer Res Treat*, 15(2), 249–256.

Stabin MG. 2007. *Radiation Protection and Dosimetry: An Introduction to Health Physics*. Springer, New York, NY.

Turner JE, Hamm RN, Ritchie RH, and Bolch WE. 1994. Monte Carlo track-structure calculations for aqueous solutions containing biomolecules. *Basic Life Sci*, 63, 155–166.

Wright HA, Magee JL, Hamm RN, Chatterjee A, Turner JE, and Klots CE. 1985. Calculations of physical and chemical reactions produced in irradiated water containing DNA. *Radiat Prot Dosim*, 13, 133–136.

# Chapter 10

# Future Needs and Prospects

Niels Bohr is quoted as saying "Prediction is very difficult, especially if it's about the future." What will be the future of the practice of internal dosimetry in nuclear medicine? The author feels significant trepidation in attempting to address this subject, but a feeble attempt will be made to address a few relevant areas.

## Pharmaceuticals

An excellent overview of the use of diagnostic radiopharmaceuticals in the United States over a 50-year period was provided by Drozdovitch et al. (2015). The development of new diagnostic radiopharmaceuticals is an ongoing endeavor. One may note from the tables of Drozdovitch et al. the significant shift toward $^{99m}$Tc radiopharmaceuticals over the years, as this nuclide has very attractive properties in terms of compound labeling and gamma camera imaging. However, the table does not reflect the temporary regression to the use of "older" radiopharmaceuticals during the $^{99m}$Tc shortage (Small et al. 2013). Since the approval of $^{18}$FDG for reimbursement, the growth of radiopharmaceuticals using this and other positron emitting agents has been strong. Graham (2015) reports that there were 1.85 million positron emission tomography/computed tomography (PET/CT) studies done in 2011, with $^{18}$FDG representing 95% of all procedures. FDG is a nonspecific marker of cellular metabolism and may be used in several applications. It is most widely used to detect neoplastic tissue and to monitor the progress of therapy. Graham notes that a number of prostate cancer imaging agents employing $^{18}$F and $^{11}$C are being developed. The various somatostatin receptor ligands (DOTATOC, DOTATATE, DOTANOC) labeled with $^{68}$Ga for diagnosis and $^{90}$Y, $^{188}$Re, and others for therapy have grown into a significant part of the market in recent years. Short-lived nuclides for myocardial perfusion imaging such as $^{15}$H$_2$O and $^{13}$NH$_3$ require an in-house cyclotron, which is not available to all centers. The generator-produced $^{82}$Rb is relatively expensive, and has issues with $^{82}$Sr and $^{85}$Sr breakthrough. Graham also notes the $^{18}$F-labeled agent *flurpridaz*, which may prove to be a useful alternative.

There are many useful imaging agents for almost any organ and tissue of the body; nonetheless, every year at technical meetings, presentations are given on a myriad of potential new agents. The cost of bringing them to market and the issue of their effectiveness cause many agents to be investigated, but never developed into commercial products. The investigations will clearly continue indefinitely. New therapeutic agents are of particular interest; the key issue in their effectiveness—as will be briefly discussed in the following text, and as was extensively discussed in the previous chapter—is clinical acceptance of dosimetry to optimize therapies for individual patients.

## Technology for Activity Quantification

Quantification of administered activity to patients is generally done using a *dose calibrator*, which is an ionization chamber with a well environment that syringes, vials, and other objects may be inserted (Figure 10.1). The nuclear medicine gamma camera—or *Anger* camera (Anger 1957)—remains the workhorse of nuclear medicine imaging (Figure 10.2). The dose calibrator is a very reliable and easy-to-use device. The Anger camera uses sodium iodide (NaI) crystal technology. The energy resolution is poorer than other detector materials, such as germanium semiconductors, but is adequate for medical imaging. However, a new technology, employing cadmium zinc telluride (CZT, or *cad-zinc-telluride*) semiconductor detectors was recently developed for cardiac tomographic imaging (Figure 10.3). This ingenious device provides tomographic images of the heart in very short times by oscillating vertical detector arrays. CZT has a much better energy resolution and

**Figure 10.1    Dose calibrator. (From Capintec, Inc., http://www. capintec.com/product/crc-pc-smart-chamber.)**

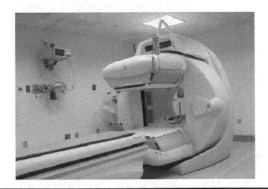

**Figure 10.2    Nuclear medicine gamma camera. (From Agetec, Inc., https://www.alibaba.com/product-detail/NM-Gamma-Camera_ 113441366/showimage.html)**

Figure 10.3  Cadmium zinc telluride–based cardiac imaging system. (From Spectrum Dynamics, http://www.spectrum-dynamics.com/products/d-spect.html)

detection efficiency than NaI; it would be an important step forward if general gamma cameras could be developed using this technology, as imaging times would be shorter, and, if used for dosimetry, the improved energy resolution would be an advantage. PET/CT devices may employ a number of scintillation detector types, including bismuth germanium oxide (BGO), gadolinium oxyorthosilicate (GOS), or lutetium oxyorthosilicate (LOS) (Figure 10.4). Research continues into detector types for PET imaging.

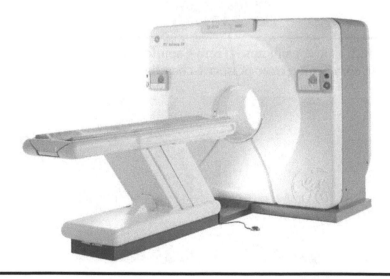

Figure 10.4  Positron emission tomography (PET) scanner (From Portal da Enfermagem, http://portaldaenfermagem.com.br/admportal/Galerias/Galeria_Interna/417201493643AM.png)

## Dosimetry Models

As noted in Chapter 2, models for dose calculations have evolved from simple spherical models to sophisticated and highly realistic body models based on medical images of actual human patients (e.g., Figure 10.5). As noted in Chapter 3, image-based models of some animal species have also been developed, to calculate doses to animals, if desired. It is difficult to say if these highly sophisticated models represent an *end point* in model development, but it is also difficult to see how much better models of the body could be created. As discussed in Chapter 4, the development of bone and marrow models has been ongoing for decades, but without reaching a satisfactory endpoint. This research will clearly continue for many years.

## Clinical Acceptance of Dosimetry

Chapter 8 gave an extensive argument for the use of patient-individualized dosimetry in nuclear medicine therapy,

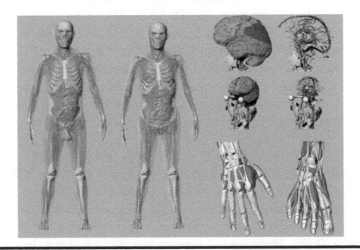

**Figure 10.5    4D adult male and female XCAT phantoms. The phantoms are highly detailed, containing thousands of defined structures and include parameterized models for the cardiac and respiratory motions. (From Segars WP et al., 2010, *Med Phys*, 37, 4902–4915.)**

and refuted all the arguments that are often made for not performing this important procedure. Resistance to performing patient-individualized dosimetry remains strong, but attitudes are changing, certainly at least in the European Union. From a logical perspective, there is absolutely no reason to treat nuclear medicine patients with a lower standard of care than cancer patients receiving external beam therapy for cancer (Stabin 2008). It appears that the new European Union Directive will mandate this practice by 2018; one can hope that this will be an example to regulators in the United States to act in kind.

# Bibliography

Anger H. 1957. A new instrument for mapping gamma-ray emitters. *Biology and Medicine Quarterly Report UCRL*, 3653, 38. University of California Radiation Laboratory, Berkeley.

Drozdovitch V, Brill AB, Callahan RJ, Clanton JA, DePietro A, Goldsmith SJ, Greenspan BS, et al. 2015. Use of radiopharmaceuticals in diagnostic nuclear medicine in the United States: 1960–2010. *Health Phys*, 108(5), 520–537.

Graham MM. 2015. *PET Radiopharmaceuticals in Development. Advance Healthcare Network Executive Insight.* http://healthcare-executive-insight.advanceweb.com/Web-Extras/Online-Extras/PET-Radiopharmaceuticals-in-Development.aspx

Segars WP, Sturgeon G, Mendonca S, Grimes J, and Tsui BM. 2010. 4D XCAT phantom for multimodality imaging research. *Med Phys*, 37(9), 4902–4915.

Small GR, Ruddy TD, Simion O, Alam M, Fuller L, Chen L, Beanlands RS, and Chow BJ. 2013. Lessons from the Tc-99m shortage: implications of substituting Tl-201 for Tc-99m single-photon emission computed tomography. *Circ Cardiovasc Imaging*, 6(5), 683–691.

Stabin MG. 2008. The case for patient-specific dosimetry in radionuclide therapy. *Cancer Biother Radio*, 23(3), 273–284.

# Index

Note: Page numbers in *italics* indicate figures and tables.

Milton Keynes UK
Ingram Content Group UK Ltd.
UKHW040059071024
449327UK00019B/663